An Elephant called Eleanor

An Elephan

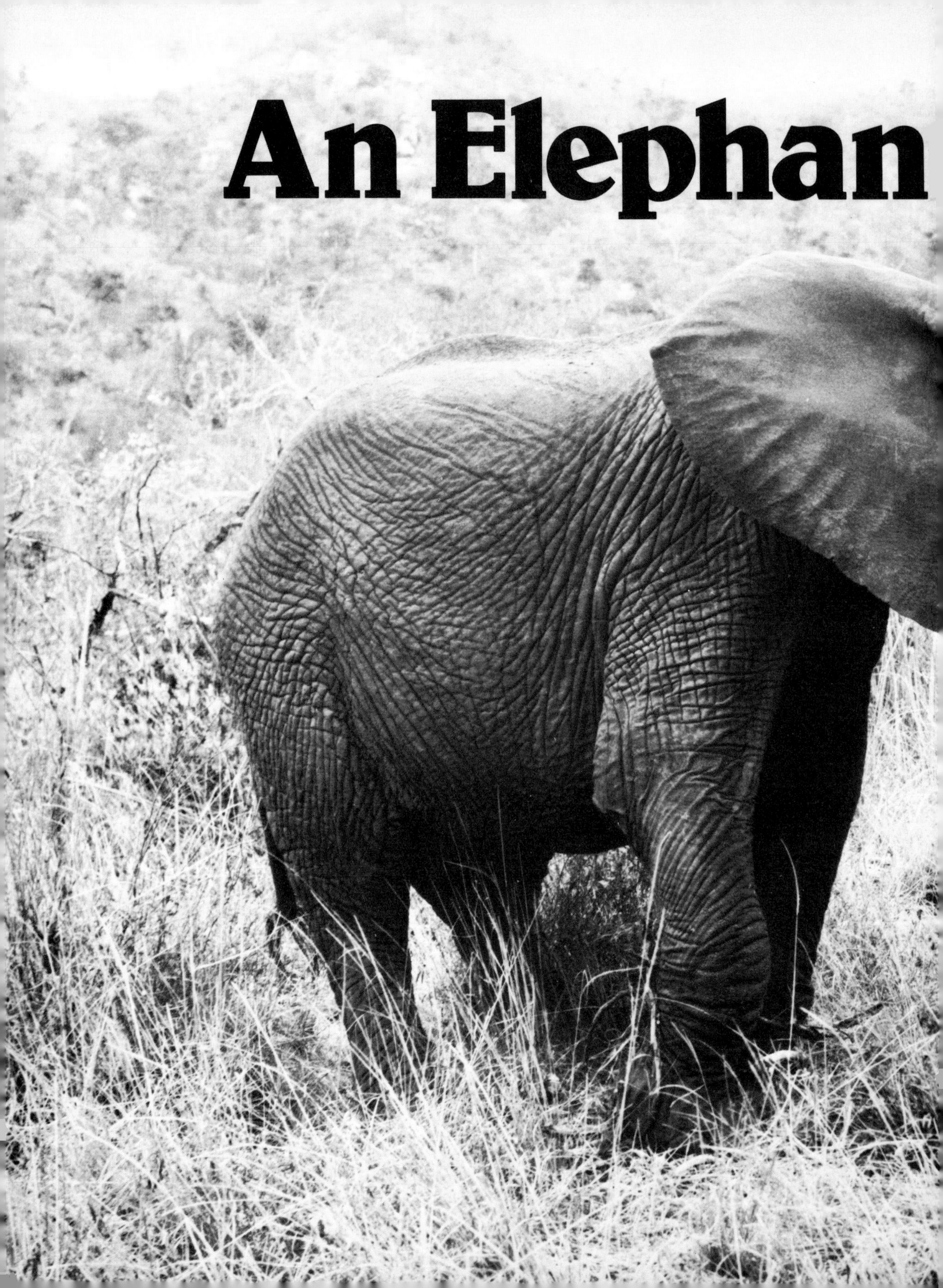

called Eleanor

Daphne Sheldrick

J. M. DENT & SONS LIMITED
London Toronto Melbourne

First published 198
Printed and bound in Great Britain by
Butler and Tanner Ltd
Frome and London
for J. M. Dent & Sons Limited
Aldine House, Welbeck Street, London
This book is set in 12/14pt VIP Plantin

British Library Cataloguing in Publication Data

Sheldrick, Daphne
An elephant called Eleanor.
1. African elephant 2. Mammals – Kenya – Tsavo
National Park 3. Tsavo National Park, Kenya
I. Title
599′.61 QL737.P98

ISBN 0–460–06900–4

Contents

Introduction

I am wondering how to begin this story about Eleanor. Because she is so different from any other animal, and so exceptional in a very special sort of way, perhaps her story should have a beginning that is different from others, and rather unusual too.

Eleanor stands nearly ten feet tall and weighs just under four tons. When I stand beside her the top of my head just reaches the top of her foreleg. Her skin is like upholstered leather; a patchwork quilt that is prickly to feel because it is covered in short, stiff hairs. Mostly it is coarse and grainy to the touch, yet in some places it is pliable and spongy, especially where it hangs in loose, baggy 'pants' above her back legs; behind her ears it is soft and cool and as smooth as silk.

Her ears are one of Eleanor's most outstanding features. They are enormous. Round and wafery, they fan slowly ever backwards and forwards, stirring the hot still air of her arid homeland into a gentle cooling breeze, and flapping softly as they slap against her flanks. Hers are probably the most sensitive ears in the entire animal kingdom: huge sounding boards that trap the faintest whisper borne on a breath of wilderness wind, enabling her to unravel the symphony of different sounds that reach her and decipher their significance and meaning. For her world is different from ours, and one where every sound has a meaning.

Her legs are tall, straight columns that easily support her massive bulk. Her feet, tipped with shiny round toenails, have cushioned soles that, despite her great size, enable her to move with uncanny silence through the bush covering the ground swiftly and easily in measured strides that appear deceptively deliberate and slow.

Two gleaming cream coloured tusks protrude from her head. Between them hangs her most unique feature – her long, rubbery trunk, its twin projections at the tip serving her very ably as a hand. Her trunk fulfils a multitude of functions besides that of breathing. It is flexible and strong, and like an arm, immensely powerful and supple. It is also extremely sensitive to scent, so can tell her more about her world than either her ears or her eyes. Raising it like a periscope, she can reach high in the air to test the wilderness wind. In this way she can determine by scent who moves beyond the field of her vision, and whether they be friend or foe. She is therefore usually forewarned of danger, and can easily trace a missing friend. Eleanor can determine where the rain has fallen in the hot thirstland that is her home; even if it be far away, she can tell where dry depressions in the landscape have been transformed overnight into life-giving drinking pools by rain, and where the bush will soon be tinged with green. She uses her trunk to fondle and feel, punish or help, rub an itch or remove a thorn, but mainly to select by smell what she should or should not eat, and convey both food and water to her mouth without having to lower her head, as do most other animals. With it she can pluck a tender shoot from the topmost branch of a tall tree, or delicately pick up a single fallen leaf without even having to look; and she can as easily break down a branch in order to peel off its bark as fell a tree if the mood so takes her. With her trunk she showers water over her body in a cooling spray during the heat of noon, or powders herself at bath time with puffs of blown dust. Her trunk also adds a very distinctive trumpet tone to her voice when she calls, making it quite different from any other voice in the wilderness.

Soft brown eyes fringed with lashes as long as a hand gaze downwards calmly and steadfastly with a deep intelligence mirrored in their tawny depths; haunting eyes of an animal which thinks and understands. As an elephant, Eleanor sees things that we never see, and hears sounds we never hear. She also knows of things that we will never know, and can predict events important to her with a mysterious certainty we would call 'eerie' long before they have even happened. For she has been endowed by nature with that mysterious sixth sense we know of only as instinct which gives her the perceptive ability denied us when we took a different turning in the road of life.

Instinct is that built-in memory with which most animals are born, which brings them a mystical wisdom few humans can understand, and dictates automatically the most important decisions of their lives. Eleanor is therefore well equipped to live with confidence in a wilderness world which we would find extremely hostile and frightening. Yet, unlike others of her

Eleanor stands nearly ten feet tall
Photo: Gerard Vienne

kind, she moves with equal confidence in a world shunned by all things wild – the world of men. She is an elephant of two worlds, and this is what makes her so different from all other elephants. Although her loyalties are divided, she remains faithful to both.

Her home is a place called Tsavo, one of Africa's largest and now most famous National Parks, positioned midway between Kenya's capital city of Nairobi and its port of Mombasa. Because it is roughly the size of Wales, Tsavo is a giant among the world's National Parks. It has been our home for close on thirty years, for David, my husband, was its first Warden and to him fell the giant task of 'taming the nyika' and creating from virgin no-man's-land a tourist attraction that would beckon rather than repel. For Tsavo has always been a naturally forbidding piece of country. No people ever lived there permanently because it is too arid due to lack of rain, and only two permanent rivers serve that huge thirstland throughout the long dry season. Tsetse-fly infest the scrubby bush, bringing sickness and death to domestic animals. Before we came to live there, very few people had ever penetrated the area, save an occasional early explorer or professional hunter in search of a good trophy, and bands of tribesmen from the Waliangulu people, who hunted the elephants with giant bows and poison-tipped arrows. No roads scarred its surface save those made by the elephants, which spidered through the bush in a filigree of fingers linking feeding grounds to watering places, skirting swamps and penetrating thickets; an intricate network of trails carved with skill to follow the best alignment over difficult country and used by all the denizens of the 'nyika'.

The term 'nyika' means much more than just 'wilderness' or 'bush'. It means both, but in a more dangerous and mysterious sense, adding a touch of fear and awe. Here, for most of the year, silvery trees barbed with cruel thorns thrust their stark, leafless branches to the skies as though in supplication for the rain that always seems reluctant to come. The burning sun beats mercilessly down from a brassy sky, parching the earth to its customary reddish brown. Like an interminable sea of grey, the nyika sprawls from one horizon to the next, and beyond, with a flat monotony relieved only by one long plateau known as the Yatta that runs through the Park like a backbone, and a few other rocky outcrops brooding in isolation over the pervading flatness. Here, time is measured only by the daily arc of the sun across the heavens, and the changing shape of the African moon, in which long, hot, dry seasons of hardship and privation give way turbulently to short stormy seasons of plenty in an endless seesaw that determines the destiny of those who live there.

Tsavo and elephants have long been synonymous, so it is perhaps especially fitting that the story of Eleanor should have Tsavo as its background. In this arid land, the elephant is the provider. Only he is equipped to tunnel into the dry riverbeds and expose underground water so that it becomes accessible to other less well-equipped animals. He can even create new waterholes by rolling in shallow depressions and gradually turning them into sizeable pools by carrying away with him huge quantities of earth plastered on his body as a mudpack. The puddling action of those huge, round feet helps to seal the bottom of the waterpans, enabling them to capture and hold precious rainwater long into the dry season. The same puddling action of those great feet raises the water table in sandy luggas – dry watercourses – and in this way tiny streams are coaxed to flow where they never flowed before, and new springs are born in a thirsty land.

By breaking up dense unproductive woodlands, the elephant can create conditions that are more favourable for many other creatures, where grasses can take root, protected in their infancy by a barrier of fallen trash above. In felling trees and pulling down the branches, the elephant provides food for others at a lower level, as well as bringing about the release of nutrients and minerals that have long been locked in wood and are thus recycled within the habitat and made available to other plants and animals. The removal of an old tree often means that seedlings struggling in its shadow become better established, and by depositing copious quantities of dung far and wide in his travels, which is then carried beneath the soil by beetles and ants, the elephant contributes greatly to the fertility of the land.

Wallowing in the mud is delightfully cooling

He destroys trees but plants others, carrying the seeds he has eaten in his great body vast distances to be deposited elsewhere in his dung. Nature's bulldozer, earth remover and gardener, the elephant has long dominated Tsavo; and elephants by their presence, have influenced the lives of all the other creatures of the nyika for countless generations.

In fact, whenever I think of Tsavo, I always think of it as a land of giants; gigantic herds of gentle giants, with gleaming ivory tusks sweeping almost to the ground; giant open spaces washed with sun; giant baobab trees that dominate the nyika like grotesque upside down relics of another world; giant contrasts and giant problems. Many people would perhaps call Tsavo an uninviting land, because it guards its secrets closely; but living there one becomes enmeshed in its subtle spell and finds instead a land of endless enchantment and magic, the more appealing for its wild and forbidding aspect. To us it became a land that we loved with all our hearts.

A National Park is an area that has been set aside by the Government exclusively for all things wild, so that in a world that is rapidly becoming engulfed in concrete, wild creatures can have a corner in which to live out their lives in their own particular way, as they have done since the dawn of time. By ensuring their survival, man fulfils the obligation vested in him in Genesis when God said 'Into Thine Hands are They Given'. But a National Park serves another very important function. It is a place where busy people from an overcrowded world can forget the hustle and bustle of their artificial lives and refresh their souls by appeasing a deeply felt primordial yearning for a more natural way of life. There is peace in the solitude of a wild place that stirs the very chords of our being, that heals and refreshes and puts things back in the correct perspective. Here, all trivial cares fade before the basic one of survival, and here one can rediscover oneself.

By the time Eleanor joined us, a little over a decade had passed since those early days when David lived in a lonely tent and pioneered the nyika on foot to uncover its secrets. Now entrance gates, proudly displaying the emblem of the National Parks – that ancient pachyderm, the rhino – were the points through which visitors from many lands constantly passed to savour the sights of nature. Roads had been built to skirt the favourite haunts of animals, many simply superimposed on those age-old elephant trails that could not be bettered by man, and in this way the once remote and forbidding nyika was brought within reach of ordinary people. Tourist lodges stood in scenic settings, dams and boreholes provided additional water for the wild inmates of the Park, and the headquarters, nerve centre for all activities, was built on the slopes of a low hill shaped like a recumbent elephant, known as the Hill of the Bees. It overlooked the brooding land it served, which stretched in a grey blanket to where the heavens dipped to merge with the horizon in a hazy hue of mauve.

Our house was sited near the Park headquarters, and David built it large and spacious so that it would be airy and cool in a land where every breeze was welcome. He landscaped the garden with the skill of a born naturalist so that it would be in keeping with its wild surroundings. No barrier or fence segregated it from the wilderness – the one merged into the other. We wanted it that way because we had always been accustomed to Africa's open spaces.

Imagine a place of tropical beauty, with an artificial stream tinkling over tumbled boulders to the accompaniment of a cooling, gurgling sound before splashing over a tiny waterfall into a pool where goldfish and guppies darted beneath the broad, flat waterlily leaves. Tall trees covered in showers of brilliant bougainvillea blossom cast a cooling shade across green lawns, where at a lower level a little pathway threaded its way through banks of bright flowers. Exotic, gaudy shrubs lent tropical flavour, giant boulders dumped at random, provided ruggedness, and terraces faced with the red rock of Tsavo gave rustic beauty to the setting.

To step into our garden was like walking into paradise, or so we thought, for not only was it unexpectedly beautiful and cool in a setting that was hot and harsh and arid, but also it was suffused by a special atmosphere of contentment and peace, for here a truce had been arranged between man and beast. Encouraged by years of gentle coaxing, many wild creatures had shed their inherent fear of man to venture into our garden and relax in cool surroundings. A flock of over a hundred beautiful vulturine guineafowl, with powder-blue breasts distinctively striped, wandered in and out as tame as any domestic chickens. Tiny dikdiks, smallest of all African antelopes, strolled unconcernedly about the flower-beds. Ground squirrels and tree squirrels scampered busily about their business, watched by orange headed agama lizards, which stood sentry on the boulders and surveyed the scene with bobbing heads. Bright yellow weaverbirds, and many other species too, came to the bird bath to drink. They chattered incessantly, filling the air with song, as they built their nests and rested in the trees.

At night, larger visitors often called, not all of them always quite so welcome. An old bull giraffe came regularly to prune the tree on the lawn with loud chomping sounds that disturbed our sleep. That would not really have mattered had he not made almost a habit of getting himself stuck in the pergola, carting it off with him and dumping it a long way down the road. A group of bachelor buffalo bulls came regularly to crop the grass of the lawn. That, too, would not have mattered had they not torn it out by its roots. Sometimes the moonlight streaming in through our bedroom window would be momentarily blocked by a huge shape as an elephant slipped silently by like a phantom, after having feasted on the banana patch below the lily pond. A wild rhino often came to partake of a scarlet shrub he particularly liked, but in order to do so, always stood on the septic tank with disastrous results. And sometimes the lions arrived. They roared and roared until the windows rattled on their hinges, and the baby animal orphans in their stables abutting the side of the house became nervous and restive. At such times, a sleepy voice from the next room would invariably chirp up in a disgruntled way, saying, 'Daddy, please chase those lions away – we can't get to sleep!' The noises of an African night were as commonplace to our daughters, Jill and Angela, as were the sounds of traffic to a Londoner.

The Orphans

The life of a Park Warden and his family is always closely linked with the lives of animals, for he is basically responsible for the care and protection of the wild animals within the sanctuary. Being a Park Warden, in Kenya at any rate, entails much more than that: he has to be architect, designer and builder, administrator, accountant and leader of men, road-maker, bridge-erector and surveyor, pilot and driver of many other machines and, most important of all, a born naturalist with that deep understanding and feeling for nature that brings with it a natural wisdom bordering on instinct. It is vital too, to be a good strategist and policeman, in order to outwit and thwart those who would plunder and destroy for personal gain.

Poaching has always been a menace in Tsavo, for Tsavo harbours animals carrying trophies that are coveted by civilized man. Furthermore, it is ringed by primitive tribes with hunting in their blood. Elephants are killed for their ivory tusks which are sold and converted into billiard balls, piano keys, carvings and trinkets. Leopards and other wild cats are killed for their beautiful skins which are made into furs for wealthy women; and the rhino simply for its fibrous horn which holds deep mystical significance for eastern peoples.

No skill is required to kill an elephant or rhino with a poisoned arrow, or for that matter, to trap a leopard, except perhaps an ability to pull a king-sized bow or set the trap. Nor does it call for particular courage or daring, for all a poacher need do is wait in ambush in a safe place – up a tree beside a well-used trail, or at a favourite watering place – and when the animal or animals come, as sooner or later they must, loose off as many arrows as possible into a herd in the hope that at least one will find its mark. Being detachable, the shaft of the arrow falls away on impact, leaving the deadly, razor-sharp, poison-smeared metal head to travel on alone and bury itself deep in the animal's flesh. It makes no difference where the arrow lodges: whether in the trunk, the body itself, or even in a leg or foot the result will always be the same, for the poison is quickly absorbed into the bloodstream and carried round the body to do its deadly work. Once an animal has been hit by an arrow and the poison absorbed, it is as good as dead, for there is no known antidote to that deadly tar-like substance made from the boiled bark of the *Acokanthera* tree. Sometimes, if the poison is strong and fresh, death comes fairly swiftly, perhaps within three or four days in the case of an elephant, but providence is not always quite so merciful. More usually death comes slowly and agonizingly over a period of weeks or even months, gradually poisoning and rotting the giant body until the flesh falls away in great suppurating hunks. There have been elephants in Tsavo, whose lower legs have fallen off as a result of arrow wounds, and others whose great bodies were so rotted that they burst like over-ripe fruit when a merciful bullet eventually brought their suffering to an end.

Because of the poaching, Tsavo was perhaps a sadder place than most, although the wilderness is always a place where tragedies are every-day occurrences. For predators must kill in order to live, and the fate of an individual animal is insignificant when gauged against the welfare of the whole. Nature manipulates her minions with a cold and brutal impartiality: she grants no privileges, permits few mistakes and rigidly prunes the wild populations so that only the best survive, and to ensure that the natural balance is always maintained. Mothers lose babies and babies their mothers, but occasionally, in the course of duty, someone happened upon that age-old tragedy of a baby animal, terrified and alone, which has lost its mother either to poachers or to predators, and is itself destined to die because it cannot survive on its own.

Those who witnessed such tragedies more often than most were the National Park Rangers, a special force of thirty men known as the Field Force, who served as David's eyes and ears. Their task was constantly to patrol the Park on foot in bands of ten. As they were equipped with radio they could report back to headquarters on all they saw and found. Their main job was to track down and capture those merciless poachers. In order to do this, David had to find some rather special people for his Field Force, for not everyone can read the nyika like a book, piecing together a sequence of events from a few tell-tale marks on the ground – a footprint in soft earth perhaps, a twisted twig or fallen leaf, or even a spot of blood brushed lightly against a grass stem in passing. Such people, who were sufficiently skilled in bushcraft to be a match for the poachers, were found only among the wild nomadic tribes of the Northern Frontier region of Kenya, and it was from there that the Field Force Rangers came. Although they could neither read nor write, nor speak any generally understood language when they arrived, they nevertheless had the basic qualities essential for the specialized work they would be expected to undertake.

Having been trained, and welded into a disciplined cohesive unit, they were sent out into the nyika, where they were constantly on the move, searching that huge Park for poachers. And, because the care of wild animals was his business, and he was a compassionate person, David gave instructions that they were to bring back any victims of circumstance they happened upon in their travels which were too young to survive without a mother. It was these animals, orphaned by misfortune and spared by the compassion of man, which came to be known as the orphans of Tsavo, and shared our lives and our home.

The orphans arrived unexpectedly in many different shapes and sizes, to be housed and cared for in the nursery stables adjoining the house. Once they could be trusted not to run away, they were allowed free run of the garden. Those of the smaller, gentler variety, who could live in the garden without wrecking it, became garden orphans. They lay around the lawn, happy to be close to us, but free to go and come as they wished. The afternoons were always set aside for a walk, which was the highlight of the garden orphans' day – and important, too, for it taught them about the wilderness. During the walk there were always exciting scents and sounds to investigate, plants and shrubs to sample, puddles and mud in which to slosh and wallow. The afternoon walk was a ritual that we were never allowed to escape, and the rattle of the teacups was the signal for those qualified as garden orphans to line up along the front steps and remind us that it was time for the daily walk.

Animals of the large variety, were only garden orphans when young, for one cannot very well have an elephant or a rhino, or even a buffalo, careering around the garden for too long. It was in any event necessary for them to spend as much time as

The afternoon walk was the highlight of the orphans' day
Photo: Sue Casson

possible feeding in the nyika to appease their huge appetites. As soon as they were weaned, therefore, they graduated to the 'Big Orphans' herd where they were housed in stout stockades behind our house, which David had constructed of steel girders and railway lines sunk in concrete, to make them elephant-proof. They were also surrounded by a lion-proof perimeter fence designed to keep at bay the rangy Tsavo lions which were seasoned killers of the larger baby animals.

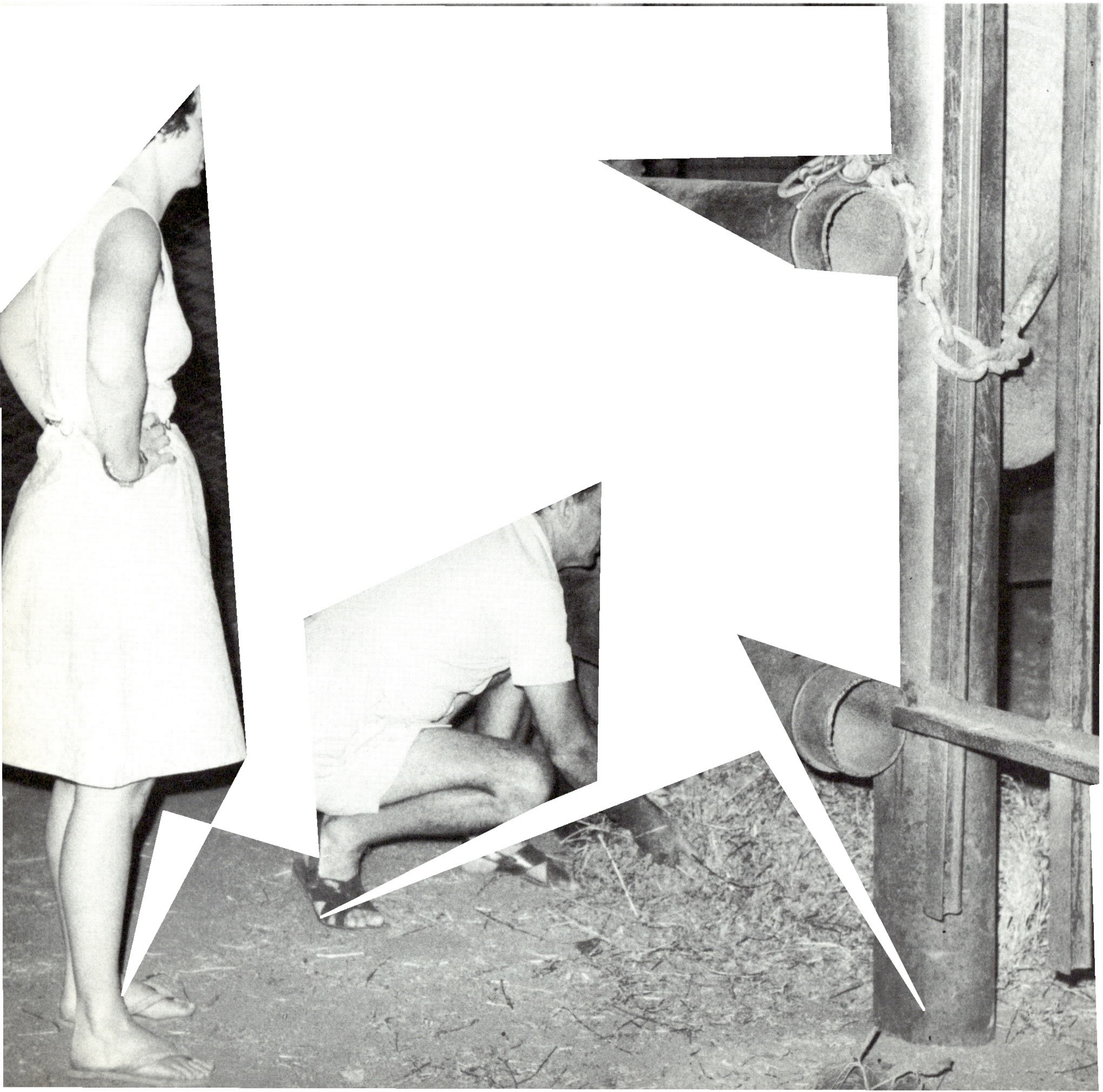

The big orphans were housed in elephant-proof stockades *Photo: Jeannie MacKinnon*

Having taken the orphans through the nursery phase of their lives, I handed them over to Ali, for it was under his eagle eye that they now came. Ali was one of those fearless tribesmen from the Northern Frontier, accustomed to herding animals and wandering in the nyika. He handled his rather unusual wild herd with the same natural aplomb he had when herding goats back in his homeland. It was his responsibility to accompany the Big Orphans, whichever they happened to be, when they were let out of their stockades first thing in the morning. His main task was to ensure that they came to no harm from lions and other animals during their wanderings in the bush. He had also to dissuade them from doing anything they should not do, such as ripping up the water pipes on their way past the Rangers' houses, vandalizing the headquarter's gardens, or raiding my orchard.

But, above all, it was necessary to have someone with them all the time to control the tourists who flocked into the Park every day, all of whom invariably wanted to see the orphans, for their fame had spread far and wide. David had to be very firm about enforcing special orphan-visiting hours for the public, otherwise, instead of feeding, the large orphans would have spent their time surrounded by an admiring throng, being petted and fed sweets from dawn to dusk. But Ali was not always as conscientious about this aspect of his duty as he should have been. He often incurred David's wrath by just 'happening' to be at the right place at the right time to intercept busloads of tourists, for he was an incorrigible show-off and loved nothing better than posing for his photograph, leaning nonchalantly against an elephant or a rhino, with his battered hat cocked at a jaunty angle and his yellow teeth bared in a forced leer. Being naturally astute, with a nomad's eye for the main chance, he found it a profitable occupation as well. The clicking of a battery of cameras would be followed by grateful hands dipping eagerly into pockets and pulling out the coins that were Ali's particular weakness. To David's intense disapproval, and though against all the rules, he made quite a useful income on the side in this way. But despite this weakness, he was a good orphan attendant and all the animals liked him.

It is not always easy to stop an animal like an elephant or a rhino doing something it really wants to do, or make it do something it really does not, but Ali seemed able to exert a good measure of control simply by his personality. Dancing up and down in front of any offender, he would waggle a flimsy twig in front of its nose, or trunk, and castigate it with a long tirade of threats he could not possibly have carried out. At no time was the twig he carried ever used for anything other than waggling, for it was forbidden to beat the orphans, and about this David was inflexible. To beat an animal, especially an animal like a rhino or an elephant, is to incur its resentment and court disaster. The obedience of such animals must be based on mutual trust and a desire to please rather than be rooted in fear. Ali understood this better than most, and was careful not to break that trust. He treated his charges rather like people; holding a tail or a trunk for support, he would amble alongside them, while keeping up a long dissertation. He would chat about the weather, of his distant homeland and his many wives and goats, roaring with laughter and approval at any unusual antic, and yelling his anger when displeased. When he spoke to them, it was as though they understood every word.

When one keeps animals, one discovers that each is an individual in its own right, with very distinctive personalities. No two, even of the same species, were ever the same; they differed as greatly as people. Being foster mother to so many animals was therefore not only a very rewarding experience but an interesting and enlightening one as well.

It is often said that the love of an animal is like the love of God, demanding nothing, yet giving all. But this is not strictly true. The real affection and deep trust of an animal is not lightly won, and is something different and much deeper and more rewarding than the cupboard love for a provider of food. It is something that has to be painstakingly earned: with gentleness of approach, compassion, understanding and endless patience. Like all worthwhile achievements, it involves some personal sacrifice as well, not only in the physical work of an endless round of bottles and constant cleaning up, but in being there at all times to provide, simply by one's presence, that vital

Holding on to a tail, Ali kept up a long conversation with his charges *Photo: Sue Casson*

companionship, security and feeling of wellbeing that only a mother can impart. Many were the times that I would have liked to be elsewhere for some special occasion, but had to remain behind because some orphan had to be fed every few hours, or happened to be unwell at the time, or because another would pine if I went. Being mother to a baby animal is a full time commitment; and because the love one feels must be sincere and come straight from the heart, there are no short cuts, for one cannot ever pretend with animals. They instinctively know the exact depth of one's feeling.

It follows too that the heartbreak that walks in the wake of failure is deeper and more sincere, for not all our efforts at raising young orphaned animals have been crowned with success. There was, moreover, one rigid rule about which David was absolutely unbending and which we, his family, had to promise to understand and always follow. It was that the animals which came to us were only on loan from the wilderness for as long as they were dependent upon us and chose to stay; that because they were born wild, they must be given the opportunity to live wild again, for this was their destiny and purpose in life, and their birthright. To nurture them in infancy and care for them during their formative years was our task: we then had to complete it by doing our best to prepare them for the wild state. This was the trust we accepted with each animal that came to us from the nyika, for being custodians of the natural world about us we knew that to deny a wild animal its freedom is to frustrate all its natural instincts, and betray the trust that it has vested in us.

This last part of our duty caused perhaps the greatest heartache, for saying goodbye is never easy, especially to loved ones, knowing, perhaps better than most, the risks attached to living free. But the pride and deep satisfaction that comes with achievement, and the joy of witnessing a hand-raised animal rediscover its natural identity and its lost world among its own kind, fulfilling its inborn basic aspirations and instincts, supersedes all else and makes it worthwhile. To stop beside a wild herd, call an animal by name, and see it move away from its wild companions and come to you with confidence and trust is reward enough. But the ultimate reward lies in attaining the abiding love of an elephant. It was this that enriched our lives in Tsavo. And the elephant that gave more than any other was Eleanor. She was so much more than an orphaned elephant – she was a trusted friend.

Strangely enough, Eleanor's life began not in Tsavo but in the arid Northern Frontier, not far from a place known as Samburu where David recruited his Field Force Rangers.

The Beginning

After twenty-two months cradled in her mother's womb in preparation for her time on earth, Eleanor arrived in a world of legs that gently nudged and urged her to her wobbly feet. She was greeted by the soft touch of many trunks feeling her all over, and the large, comforting shape of her mother overhead. All around were gathered her blood relations – perhaps her grandmother, aunts, sisters and possibly some young brothers and cousins, all of them eager to inspect and welcome her into their midst. As she tottered to her feet, supported by her mother's trunk, she stood a little over two feet tall and weighed about two hundred and thirty pounds. Because she was an infant, still covered in a soft baby fuzz, she appeared much darker and hairier than her elders and moved with the hunched, shuffling gait of all newborn elephants.

Between her mother's forelegs, she found the milk that would nourish her for the next two years of her life – very special milk, different from any other in the animal kingdom, the composition of which would change with the passage of time to suit her growing needs. When she suckled, she would use her mouth and not her trunk, which would be coiled above her head to keep it well out of the way as she fed. And all the time she would be reassured and comforted by the low rumbling of her mother, which, in elephant language, is a sound of greeting, of love, and of pleasure.

In the months that followed, she gradually became accustomed to the other sounds made by her kind: the trumpeting of an elephant that is alarmed, playful or afraid, the sharp scream of one being punished or in pain, and the ominous roar that accompanies real anger.

Most of her babyhood was spent actually underneath her mother's body or pressed as closely to it as possible, for she needed protection against lions and other predators which, given the chance, would like to dine on baby elephant. The lazy hours of noon and part of the night were a time for sleep, when she would collapse and lie flat on her side, with her tiny trunk coiled closely against her heart. Usually her older relatives slept standing up, but sometimes, when the darkness was deepest, they too might join her briefly, lying flat out on the ground.

The heat of the day is a time for sleep. Eleanor lies on the ground while an elder relative sleeps in a standing position

The silence of the night would then be disturbed by heavy breathing and the sound of gentle snores.

When the moment came for the herd to move, she would be roused by a gentle nudge from her mother's hind foot, the signal for a baby elephant to wake up and follow.

To start with, Eleanor would have found her own trunk more of a hindrance than a help, because she would always be treading on it, or tripping over it. It was nevertheless useful to be able to put it into her mouth and suck it as a baby sucks a thumb. Gradually, by experimenting and copying her elders, she began to discover its other uses, and soon learned the value of that wonderful appendage with which only elephants have been blessed.

At one year old, she stood about three feet tall. This was an important milestone in her life, for now she was no longer entirely dependent on her mother's milk, but able to supplement her diet with some of the green plants her elders ate so avidly. By now she was much more venturesome and confident, and enjoyed playing with other small elephants when they met at waterholes, trumpeting shrilly with excitement, and dashing about in mock charges, her round ears outspread like two small dinnerplates. By copying her elders, she was able to give a good imitation of a real charge, racing up to an opponent, trumpeting loudly, reversing a few paces with her head held high so that she could look down her trunk and see straight ahead, before wheeling round to try the game elsewhere.

During those periodic visits to the waterholes, she met many elephants from other herds, for such occasions are in the nature of social gatherings. She might even have met her father, though without recognizing him, for in elephant society fathers are kept apart from the herd. They prefer to be independent, wandering on their own or in the company of other males in small bachelor groups, joining the cow herds only briefly without ever really belonging. Adult males are no more than casual visitors, for it is always the females that lead the herd and form the stable component of each group. Each herd is in fact really a family, usually led by the most senior cow of the family, with her daughters and their offspring, plus perhaps some other relatives, making up the remainder of the herd. The ties binding the individual members of a family are lasting and strong, and the leader can always count on the unswerving loyalty and devotion of her family. They follow obediently wherever she chooses to take them. Should she pause to listen and test the wind for danger, they will do likewise; should she decide to wait for several hours before approaching a watering place, for fear of danger, they will stand patiently beside her, no matter how thirsty they may be. On her judgement they will depend entirely, and should disaster strike unexpectedly, removing her suddenly from their midst, they will be desolate, confused and lost. The herd wanders in disorder until the cow next in seniority takes over responsibility for decision making. She is then automatically accepted as leader.

At about the age of twelve teenage bulls usually sever their family ties, moving away to join with older bulls. Unfettered by the same responsibilities and loyalties as the cows, they generally become unruly and boisterous as they grow up, and are an irritation to the gentle females of the family, who are much more peace-loving and tolerant. The females are not therefore sorry to see them go: sometimes they will even drive them out should they linger too long and cause too much disturbance.

Growing up amongst elephants, Eleanor experienced the inherent courtesy of her kind, accepted not only by members of herds that were known to her, but by strangers as well. Everywhere she was greeted kindly, even by the strange visiting bulls she came across in the course of her wanderings. If ever she found herself in difficulties, stuck in the mud of a waterhole, for instance, or out of her depth, another elephant would come to help her, not necessarily her mother, but perhaps a strange cow or bull from another herd. Such touching concern for the plight of others, and particularly the young, is something that is one of the most endearing characteristics of elephants. One day Eleanor herself would extend this concern beyond her own kind to many other creatures too.

She would notice the respect given the biggest bulls carrying the largest tusks, for such monarchs in the community are treated with deference by all. They enjoy certain privileges, especially priority at the watering places. Advancing with that slow deliberate confidence that brooks no argument, all lesser creatures make way at their approach; but had she, as a baby, moved close to drink at the same waterhole, she would have been allowed to do so, for even the king of the community makes allowances for the young.

Thus, her early life revolved around others of her kind and, of course, around the alternating seasons which dictate the movement of the elephant herds for, like Tsavo, her homeland of Samburu was an arid area. The long dry season was the season of sorrow, for it was then that the old, the sick, the maimed and the weak lay down to die. Elephants probably grieve the loss of a loved one more deeply than any other animal for, like us, they are long-lived, have good memories, make lasting friendships, and both give and receive a great deal of love. They are also highly intelligent. Often, they will stand sadly for hours, beside the body of a loved one, fondling it, and trying to lift it to its feet. Sometimes they will pile brush and earth on top of it, as though hiding it from their sight. Occasionally they will tear the tusks from the skull, carrying them far away, or even smashing them against rocks, as though comprehending that ivory lies at the root of much of their suffering.

The elephants of Samburu, like those of Tsavo, are anchored during the long dry season to the few sources of permanent water: it is then that the poachers are most active, bringing havoc and suffering to the herds. Somewhere in Samburu, when Eleanor was just two years old, a poisoned arrowhead fired from a poacher's bow changed the course of her life. When Eleanor was found, after a night of drenching rain which marked the end of the dry season, she was standing alone on an empty sodden plain, a forlorn and dejected little elephant, with no others in sight except the inert body of what must have been her mother lying some way away with an arrowhead embedded deeply in her side.

She was more fortunate than most, for it so happened that a party of people, including the then Governor of Kenya, Sir Patrick Renison and his wife, were touring the Samburu Reserve at the time. They were being conducted by the Warden of the Mountain Parks of Kenya, Bill Woodley, who had previously worked with David in Tsavo. He was one of the few to have accompanied David in those early days when he pioneered his Park, and had been in Tsavo when the very first elephant orphans had been brought in.

Being without a mother is bad enough for any baby animal. But for a baby elephant which has always been enveloped in the love and care of many others, to find herself suddenly surrounded by men must be a particularly terrifying experience. She dashed wildly about trumpeting in terror as she tried to evade her captors, but they quickly closed in on her from all sides and pinioned her to the ground. On this occasion, her frantic cries for help brought no response, being tossed back at her in a dying echo, for neither her family nor any other elephants were near enough to hear. The rain of the previous night had released them from the bondage of their dry weather haunts, recently the scene of so much sadness, stimulating them to trek overnight far inland where the waterholes would be filling and the grass soon turning green. Eleanor's mother, too ill to follow, had remained behind to die, and although Eleanor must have longed to accompany others of her family, she could not bear to be separated from her mother.

Strong arms loaded her into a lorry, and held her fast until she was unloaded and pushed into an improvised stable on the verandah of a tourist lodge chalet. It was at this moment that she was named Eleanor in honour of Lady Renison.

Eleanor soon lost her initial terror of men, for loving hands fondled her just as loving trunks had done before, and tasty titbits were proffered which were irresistible to a little elephant that had just endured a long dry season. Preparations were soon underway for her long journey back to her new home in the Mountain National Park a hundred miles away. As soon as she could be trusted not to run away, she was allowed out of her pen in the company of an African Ranger. His task was to be with her at all times, both day and night, for it is essential that a baby elephant should never be left alone to pine. Then she was persuaded to walk up a ramp on to the back of a lorry.

Once aboard, a special treat awaited her as her reward. Short lorry rides followed. At first she was alarmed, but the drive soon became routine, and always at the end of the journey there was something special to eat as a reward. So, when the day came for the long journey to her new home, she was quite accustomed to travelling on the back of the lorry. By then she was also trusting and unafraid of humans, for her Ranger friend was with her constantly to reassure and comfort her.

Her new home in the Mountain National Parks was very different from the arid lowlands of her birth. Here mists masked the sun most of the day for months at a time, and rain often fell in an incessant icy drizzle. Frost sometimes laced the tips of the leaves and the grass stems in the mornings, and the nights were damp and bitterly cold. There were times when she was unwell as she became acclimatized to the cold climate of her highland home. Over the next eighteen months, she nevertheless thrived, and the deep trauma of her capture gradually faded from her memory as she wandered happily with her Ranger friend in the lush forests of the Aberdare Mountains.

During this time Eleanor became a film star, for a baby elephant had to be chased by the lioness *Mara*, which played the part of *Elsa*, in the film *Born Free*. Eleanor probably experienced real terror once again, for to be chased by a lioness is a very frightening experience, more especially for a baby elephant on which lions sometimes prey. Even though it was only for a film, and therefore not real, Eleanor could hardly have been expected to know that. Sometimes she was taken to local agricultural shows. Wherever she appeared, she was an instant success, for although the townsfolk had lived all their lives in the land of wild elephants, few of them had ever seen an elephant – or even a picture of one – before.

The Nairobi Agricultural Show, which was a huge annual event, was probably her most exacting public appearance, for here a sea of incredulous, staring faces gawked and gaped from dawn to dusk, and enormous crowds surged around the stand where she was being exhibited. Whenever she reached her trunk through the bars of her stockade to greet them, they squealed and recoiled in terror, and she had only to do even the most ordinary things an elephant does such as spraying water over herself, or scratching her ear with the tip of her trunk, to make the crowd roar with laughter and shout in amazement. Consequently, by the time the official opening was due to take place, Eleanor was becoming rather bored. She had learnt that her persecutors could easily be dispersed by a shower of muddy water sucked up from her mudwallow and sprayed from her trunk, so when the dignitaries arrived in all their finery to declare the Wildlife Stand officially open, they received the same treatment, much to their acute embarrassment and the delight of the spectators.

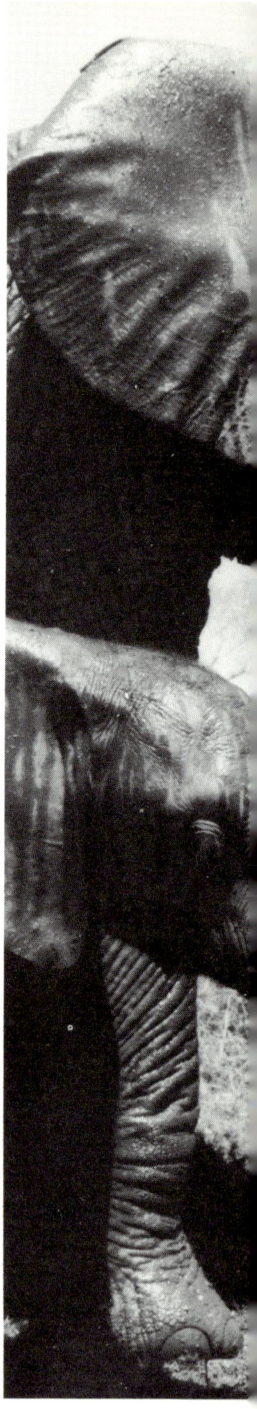

Eleanor, leader of the orphans

Never before had an elephant walked in the grand gardens of Government House, the Governor's city residence, nor, I feel sure, will one do so again. Lady Renison, after whom Eleanor was named, expressed the wish to see her, which was like being summoned by the Queen, so off to Government House Eleanor went, on the back of a lorry as usual, causing quite a stir amongst the passing city traffic. Heads swivelled and brakes screeched as astonished motorists stared in amazement to see an elephant passing by. On arrival Eleanor was greeted by the Governor and his wife in person, and was allowed to wander freely over those immaculate emerald lawns, helping herself to anything she fancied from the tempting array of exotic flowers confronting her on every side.

There followed a rather sad period in Eleanor's life, for she had made such a great impact at the Nairobi Agricultural Show that the National Park authorities insisted that she should continue to be on view to the public. She was therefore taken to the Nairobi National Park Orphanage just beyond the city limits, which was really no more than a poor imitation of a zoo. The authorities insisted that Eleanor's presence there would help publicize the Nairobi Animal Orphanage: it did, of course, but at the expense of her health. Deprived of the exercise vital to an elephant and the variety of food necessary for a balanced diet, denied the wide open spaces so essential to her psychological wellbeing and the loving individual attention she needed, the sparkle slowly faded from her eyes, and she became morose, obese and lethargic. She became just one unfortunate animal behind bars among many going through the same dreary daily routine almost mechanically, with her mind far, far away. Wide open spaces washed with sun, sparkling rain-brimmed pools, pastures green and huge grey shapes, loving rumbles and the gentle touch of tender trunks – there must have been a constant ache in her heart. She would undoubtedly have died had not Bill Woodley fought to secure from officialdom her reprieve, persuading those in authority that she must make one last journey – this time to Tsavo. There she would be among friends – animals which like her came to be known as the Orphans of Tsavo.

The Others

Rufus was so gentle I could put Angela on his back *Photo: Gerard Vienne*

Samson was the Leader of the Orphans. At twelve years of age, he was over eight feet tall with a massive frame and fair-sized tusks – an impressive teenage bull. He and a gentle female of the same age, called Fatuma, had been the first elephant orphans of Tsavo. Shortly after the Park had been established, and during a particularly harsh dry season, the Field Force Rangers had found them abandoned and had brought them back. Being a cow, Fatuma had automatically become the leader; Samson and the other animals were content to follow wherever she and Ali chose to take them. But, as she grew up, Fatuma began to feel the call of the wild and, one day, quite unexpectedly, she made her choice. She went off to join a wild herd while Samson became Leader of the Orphans.

'The herd' was constantly changing, with frequent arrivals and departures. As other animals grew up more quickly than an elephant, Samson had witnessed many changes in his time. He now inherited two three-quarter grown rhinos, three ostriches, and half a dozen buffaloes of assorted sex and size. He treated them all exactly as though they were elephants, expecting them to enjoy the same things as himself. Every day the ostriches, for instance, were dragged by their necks into the mudwallow and expected to wallow. This they obligingly condescended to do, but it resulted in their feathers being plastered to their bodies by caked mud, giving them a bedraggled appearance.

The elder of the two rhinos, Rufus, was a sloppy, lovable animal and Samson's special favourite among the orphans: he and Rufus spent long hours playing together. During these bouts, Samson usually lay down in order to make the game a little more even. Nevertheless, he easily warded off Rufus' furious onslaughts with an effortless shove of his mighty trunk, which used to drive poor Rufus almost to distraction.

The second rhino, Reudi, was not quite so good-natured, being of a more fiery temperament than Rufus. From the beginning he displayed pronounced rhino-like characteristics, and it had taken a lot of time and patience to persuade him to be accommodating to the others. This applied in particular to poor Rufus, who was usually singled out as a target for instant attack, although fighting was the very last thing Rufus ever

Like all good friends Rufus and Reudi sometimes had their differences *Photo: Gerard Vienne*

wanted to do. In time, however, Reudi gradually settled down and the two rhinos became quite good friends – for most of the time, that is. Like all friends they often had their differences which sometimes degenerated into a tremendous battle if Reudi lost his temper, which he frequently did. Then Samson would feel obliged to intervene to restore order in the ranks. There would be trumpeting, roaring, huffing and puffing, with rhinos, ostriches and buffaloes flying in all directions and Ali clinging to a branch of the nearest tree, yelling his disapproval.

Eleanor arrived late one afternoon in 1961, having travelled in her customary way on the back of a lorry. We were expecting her, of course, so the stockade next to Samson's had been prepared for her and filled with an assortment of Tsavo's most tempting shrubbery. This was cut fresh every day for the orphans to eat at night.

She walked slowly down the ramp, an extraordinarily shaped little elephant, for her tummy had become distended like a barrel as a result of the sedentary life she had led in Nairobi. Hesitantly, she extended her trunk in greeting to each of us in turn, rumbling pleasure deep in her throat, and then paused at the entrance to her stockade, obviously startled by the strange sounds emanating from the rhinos' quarters next door. Samson, in his exuberance at seeing another elephant, was trying to scale the bars of his stockade, and in this attitude probably looked very un-elephant like. After a good deal of coaxing she finally went in, and in that moment a legend was born in Tsavo. For this gentle elephant would share our lives for the next two decades, and contribute greatly to our understanding of animals, in particular to our knowledge of her kind. Looking back on those days, we always relate events connected with the orphans to Eleanor, in rather the same way that events in history are related to the reigning monarch of the period. Eleanor must have been such and such an age when so and so came or went, we say, and in this way, everything slots into place, so lasting has been her impact.

It took a little while for Eleanor to regain her health and adjust to her new surroundings, and especially to become accustomed to her two rather turbulent rhino herd-mates, of which she instantly disapproved. Being peace-loving and gentle herself, their rowdy squabbles always upset her, and she developed a strong antipathy towards rhinos generally which was to last for many years. But she adored Samson. It had been a long time since she had been with another elephant, and suddenly having one older and wiser to lean on made her feel secure and content. Samson was happy to be the subject of such adulation and treated her with gentleness.

Each morning, when the gates of the stockades were opened with a loud creak that we could hear from our bedroom, the two elephants would hurry out and lead their little band down past the orchard and house to the Rangers' houses, where they usually paused to pick up any offerings, and then into the bush in the vicinity of the Voi river. Most of the year the river itself was dry, for it was a seasonal sand watercourse that flowed only when heavy rain had fallen in the nearby Teita Hills but, as during the dry season the vegetation in its floodplain was always a little more lush than that of the nyika proper, many animals congregated there. Rocky hollows in the riverbed formed stagnant pools which were perfect for bathing hot bodies, while mudwallows left by receding floodwaters lasted long after the last rain had fallen. These formed a focal point for the animals of the nyika during the heat of noon. Elephants, rhinos and buffaloes gathered there at this time to plaster their bodies with cool mud, so Eleanor had many opportunities to meet her wild kindred.

She and Samson enjoyed mingling briefly with the wild herds of elephants. Samson, in particular, found sparring with other young bulls much more fun that pushing feeble Rufus; but events sometimes took an exciting turn if a wild rhino happened to be first ensconced in the mudwallow. This spelt trouble for, unlike elephants, wild rhinos are definitely anti-social creatures, a sentiment that was fully reciprocated by both Rufus and Reudi in their attitude towards other members of their kind occupying mudwallows. There would be a great deal of huffing and snorting and shunting about, as rhinos glowered menacingly at one another with lowered heads. Once again, either Samson or Ali would have to intervene and prompt the wild rhino to vacate the wallow so that the orphans could wallow peacefully and set about plastering themselves nicely in red mud. Tsavo is famous for its red elephants, for the red laterite soil there is the colour of red ochre, and makes the best mudwallows. Hence Tsavo elephants are always bright red, and Eleanor soon looked like a true Tsavo elephant.

Wallowing time is a time for playing, splashing, romping and rolling about. Ali had to oversee the proceedings to ensure that the ostriches were not handled too roughly or even drowned, for Samson's idea of a game with ostriches involved

Wallowing is a time for playing, splashing, romping and rolling about

Ali had to oversee the proceedings

whirling them round and round by the neck. Eleanor was fascinated by the ostriches, but was much more gentle in her approach. She liked catching hold of their bunchy tail feathers and propelling them along by pushing them from behind.

Having thoroughly plastered themselves, the orphans would then gather around Ali as he sat himself down in the shade of a nearby tree to extract his lunch from a battered old teapot in which he carried it. Grouped about him in a tight circle, they would watch every mouthful, until Ali could bear it no longer, and begrudgingly, amidst loud protests, tossed an occasional morsel over his shoulder to whichever animal could catch it. Having disposed of lunch, all the orphans would settle down for a nap, dozing away the noonday hours till the sun dipped lower in the sky and the heat became less oppressive. Stretched

The ostriches always squatted, their necks protruding like thin periscopes

out on the grass, with his hat over his eyes, Ali was sound asleep, while Samson and Eleanor both stood guard nearby, slowly fanning their huge ears as their long-lashed eyelids drooped. The ostriches always squatted down on their haunches, their necks protruding from the grass like three thin periscopes. The buffaloes lay about chewing the cud, a picture of bovine bliss, while Rufus and Reudi slept even more soundly than Ali, breathing large, contented sighs through their rubbery nostrils. This was a peaceful, silent, sleepy time of the day, when the only sounds were the singing of birds, the humming of insects, and the wet, gentle slap of elephant ears.

Afternoons were feeding time again, when the orphans fanned out and wandered slowly through the bush, taking a mouthful of grass here and plucking a leaf or a twig there. Samson liked to demonstrate his prowess at pushing down the spindly nyika trees. When that happened, the quiet munching and chomping and rustling that marked the passage of the orphans through the nyika would be interrupted by the tearing, wrenching and breaking of wood.

Ali would begin to round up his charges in the late afternoon. They would then file slowly back to the stockades, Samson striding ahead of the column with Eleanor hurrying to keep up. But on arriving opposite the side of the house, she would pause and listen carefully, for there were always sucking sounds coming from the nursery enclosure where the baby orphans would be having their evening bottle. Eleanor's curiosity about the occupants of the nursery always got the better of her. Fascinated by all babies she would deviate briefly and gaze lovingly over the fence. She would become so absorbed, that Ali would have to come back and waggle his stick in front of her trunk to remind her that she should be elsewhere.

Thus, the days passed into weeks and months, and Eleanor settled down to her daily routine. Life was largely uneventful during those early years, except for the occasional rhino problem – and until Samson began to feel that something was lacking in his life, which indeed it was. At his age, he should have been wandering the nyika with other young bulls, and practising sparring in readiness for the time when he could win a female. Often he would scent a wild herd that contained a female on heat, and he began to become increasingly restless. The oily secretion from his temporal glands that constantly stained his cheeks was evidence of the latent desires raging within him. Then, with a loud excited trumpet, he would go off in hot pursuit, leaving chaos in his wake: Ali dancing up and down in order to regain some semblance of order; Eleanor, thoroughly confused, was undecided whether to follow Samson or obey Ali's command to stay; Rufus and Reudi charged about blindly, not really knowing what was going on, often bumping into each other around a bush, giving themselves such a fright that a fight invariably followed; while ostriches and buffaloes ran hither and thither in all directions.

David had a special affinity with Samson *Photo: Bernard Kunicki*

Sometimes Samson would be absent for a month or more, and just as we were beginning to think that this time he had gone for good, he would suddenly reappear to vent his frustration on the other orphans. He began to create a problem around the headquarters, sneaking in during the dead of night to gobble up the gardens and orchard, knocking down all David's newly planted trees until the grounds resembled a battlefield, and turning on all the water taps he could lay his trunk on so that the tanks were drained dry by the morning. As if that were not enough, he took to bringing his friends along too, and whilst they were perfectly prepared to keep him company during his nefarious activities under cover of darkness, they were not nearly so brave when the staff began to emerge in the mornings for work. Not being used to fraternizing with human beings, they generally ended up scattering all the labourers and rampaging about until they could escape, while people took refuge in trees or fled in disarray. Meanwhile, Samson would simply carry on with whatever was occupying him at the time, unable to see what all the fuss was about, and obviously puzzled as to why he could not keep his friends for very long.

Eventually, David had to harden his heart. A bull elephant is not as dependable and predictable as a cow, and we could not subject the staff to unnecessary risk. So, thereafter, every time poor Samson decided to come back home, he was forcibly evicted, driven off with thunderflashes or otherwise discouraged, until he at length began to realize that he was no longer welcome. It hurt to have to treat him harshly, and it must have puzzled him to meet with hostility where before he had known only kindness. While it tugged at our heartstrings, and especially at David's – for he had a special affinity with this first elephant orphan of Tsavo – we knew that what we were doing was for the best. Besides, it was what would have happened to him in his own herd, had he been a wild elephant. We knew the break was painful for Samson: sometimes he stole up to the stockades in the dead of night, and paced round and round, longing to be allowed in.

I remember, too, how deeply touched I was when he returned unexpectedly one Sunday afternoon after a long absence, and advanced towards little Angela playing in the car park. I appeared just in time to see him towering over her, and my heart skipped a beat, fearing that he might harm her, having been harshly treated of late. But I should have known elephants better. Slowly and hesitantly he stretched out his enormous trunk. Then, gently and tenderly, he touched the top of her head, rumbling a loving greeting. When he saw me, his ears stood out guiltily because he knew he should not be there. Sadly he turned and went away, never to return.

During our travels around the Park we often used to see Samson. Many times we were sorely tempted to call him and renew that long friendship, but we realized that this would have been a retrograde step which would not be in his interests.

Back at headquarters, all Eleanor's female elephant instincts could now assert themselves, for it fell to her to assume leadership of the orphan herd. She accepted this responsibility with a maturity few human children her age could begin to match, even the unsavoury one of trying to keep the peace between Rufus and Reudi. With a shrill trumpet she would end the quarrel by forcing her way between them as Samson had done, flailing them both with her trunk before chasing them in opposite directions. Eleanor was only seven when she became the Leader of the Orphans of Tsavo.

The Sixties

For a while Eleanor was the only elephant in the orphan herd. With the passage of time the composition of her herd changed again. When the original ostriches grew up they took to attacking anyone who wore a hat so had to be banished to a remote part of the Park. Some of the larger buffaloes also returned to the wild. Their place was taken by another consignment of ostriches, while two of the occupants of the nursery graduated to Eleanor's herd. I remember only too well the trepidation that accompanied the introduction of these two newcomers. Stub, the baby rhino, was a pugnacious little character, quite different from Rufus. His nursery companion and lifelong friend was a buffalo, named Lollipa.

Knowing that Rufus would be the most tolerant of the rhinos, David called him first. Rufus approached at a brisk trot expecting a titbit without even noticing Stub who was bouncing up and down beneath his enormous chin. Certainly, Stub had spirit. With simple rhino-like logic, she had obviously convinced herself that the enemy had been successfully vanquished. This seemed to go to her head and caused her to become thoroughly provocative. It would probably have been better for her had Rufus been a little sterner, for Stub's

Samson and Fatuma at the lily pond *Photo: Bernard Kunicki*

self-assurance was ill-founded when she came to be introduced to Rufus's companion, Reudi.

Because we expected trouble from Reudi, we had deliberately steered him away from Eleanor and the other animals to facilitate the introductions, hoping that Eleanor would be in a position to contain the situation once Reudi became aware of the presence of the two newcomers. But matters did not go as planned for, believing that she had got the better of Rufus, Stub seemed prepared to attack every animal in sight. She charged in amongst Eleanor and the others, and initiated a rodeo by alternately chasing and being chased by every animal in turn, Eleanor included. Eleanor, who in any event was not very fond of rhinos, was thoroughly confused by the unexpected entry of this small apparition bent on causing strife among her brood. Reudi looked up, attracted by the commotion; slowly and ominously raising his tail he advanced at a bouncy trot right up to little Stub, who gallantly stood her ground. The next moment Stub found herself sailing through the air, propelled by the tip of Reudi's long horn. Fortunately, David rushed to the rescue, arriving at the bush in which Stub landed at exactly the same moment as Reudi and Eleanor. Eleanor obligingly dealt with Reudi, while a rather battered Stub was hurriedly hustled off to the nursery, along with poor Lollipa, who had not fared any better at the hands – or rather horns – of the other, bigger buffaloes. The introductions had to be postponed until bruises and stabs had been patched up, and then Lollipa, who had been halter-trained by Jill, was dragged out every day to meet Eleanor and her herd as they came home in the evening. Because Stub refused to be separated from Lollipa, she reluctantly had to go along as well, until finally both were accepted by Eleanor and Reudi, and the integration was complete. Once again, peace returned to the herd.

The decade of the sixties will always remain in my mind as perhaps the happiest and most relaxed period of all our years at Tsavo, for times were good both for the animals and for us. The rains came as expected and fell in more generous measure than usual. The huge elephant herds were able to live in peace and plenty, without being constantly molested by poachers.

The elephants lost their inherent fear of man and wandered over those vast, flat bushlands, opening up the dense thickets, wallowing in the rain-filled waterholes and generally enacting their role according to the laws of nature. Blessed with good seasons, the grass grew tall, shimmering and rippling in the breeze. Springs appeared where there had been none before; streams flowed where previously there had been only dry luggas. And the smaller nyika dwellers multiplied and gathered together into sizeable herds because, thanks to the elephants, the country was now more open.

For us, the sixties was a period of learning, when David's orderly, scientific mind could satisfy his restless quest for knowledge, and in this Eleanor played a vital role. There was time now to walk with her in the nyika, to watch and learn from her the habits of her mysterious kin and better understand their important role in nature.

As the sun rose over the distant horizon in a fiery, scarlet ball, tinting the heavens pink, we would often go to the stockades, before setting out with Eleanor and her herd into the nyika to collect samples of the plants they chose to eat or simply to watch and observe them under natural conditions.

Pressed beside her, we found that it was possible to walk in amongst completely wild elephants and be able to study them at close quarters without disturbing them in any way.

To walk in total trust with an elephant among a herd of wild elephants is an exciting and very special experience, and one that I shall remember and treasure all my life. We knew that Eleanor would shield us and would never lead us into trouble. Certainly, when we were with her she seemed very conscious of the need to hide us, for she always positioned herself between the wild elephants and us, steering us where she knew we would be in no danger. In her presence there was an unprecedented opportunity to study so many aspects of nature. We watched the parasitic flies that settled on her and the rhino; and saw how different animals reacted to the scents and the sounds of the nyika and how they behaved towards strange animals and each other. When the sun began to sink below the outline of the distant western hills, we would wend our way homewards, not forgetting to reward Eleanor with a grapefruit from the orchard.

Back at the stockades we acquired an insight into the depth of Eleanor's mysterious mind. She was very clever. She could, for instance, open the latch to the food store, lifting the bolt and flipping it across with a quick movement of the tip of her trunk, before pulling on the door to open it. Ali meanwhile was hopping up and down, heaving with all his might in the opposite direction to try and keep it shut in order to prevent her consuming everything inside. She could easily turn on a tap, using her foot and her trunk in clever conjunction with one another, deftly twisting it until the water gushed into her upturned trunk. If the hosepipe was attached, she would simply raise it to her mouth and hasten the drinking process by pouring it straight down her throat. As darkness fell, we would

Eleanor drinks direct from the hosepipe

bid her good night, saying: 'Shake hands, Eleanor!' Slowly and gently, up would come one massive foreleg until it rested on the crossbars of her pen to be patted and caressed and held.

'Put down your head, Eleanor,' and down would come that enormous forehead until it could be reached and rubbed, while low rumbles assured us of her affection, and the tip of a wet trunk snaked all over us, fondling and feeling and smelling us tenderly.

Once again there were changes in the composition of her herd, for it was finally time for Rufus and Reudi, now fully grown, to leave the orphans and return to a natural existence. It gives one a very special feeling of pride and satisfaction to look on an enormous animal like a rhino, and know that but for us it would have been a heap of bleached bones in the wilderness. We would cast our minds back ten years remembering this same magnificent beast as it was when it came to us as a soft, blunt-nosed, almost blind, dependent small creature, only eighteen inches high and in need of as much care as a human baby. Now, because we had succeeded, it still lived and breathed and stood on the threshold of a new and exciting life. I have known this thrill many times, but somehow it is always more intense with a difficult animal like a rhino.

Meanwhile Stub, the most recent addition to Eleanor's herd, had grown up a lot, as had her colleague, Lollipa, who was turning into a very special buffalo. The nursery had also filled again – I could hardly believe it – with two more rhinos. Inwardly I groaned, for rhinos are not easy animals to raise; neither are they always easy to manage when grown, as was being proved by Stub.

One of the newcomers was already about three years old when brought to us, her mother having been shot on the neighbouring sisal estate, and was therefore in a very aggressive state of mind. She expended the last ounce of her strength pounding the door of her stable and anything else that caught her eye. This prompted us to call her 'Stroppie'. Eventually, she collapsed in a coma, and for the next ten days I battled for her life. She received endless injections, until the soft skin behind her ears was like a pincushion. Every morsel of food had

Lollipa and Stub with Angela *Photo: Gerard Vienne*

to be introduced into her mouth by hand. Hours were spent simply coaxing her to make the effort to want to live. All these efforts would have been fruitless, I am sure, had it not been for the timely arrival of Punda.

Punda was a zebra foal which became separated from its mother and attached itself to a minibus painted with imitation zebra-like stripes, and having followed the bus for several miles, ended up being deposited at our house. Despite his good looks, Punda was born a rascal. He was mischievous, persistent and stubborn, but very endearing. In the years ahead he became a dominant force in Eleanor's herd, though a disruptive one as well. But we were not of course to know this as we steered him towards the nursery where poor Stroppie was hovering between this world and the next.

Punda became aware of Stroppie stretched out on her sick-bed of hay as soon as he entered the enclosure, and a wicked gleam crept into his eye. Without further ado, he gripped her button-like horn between his teeth and began to tug. The indignity of this was too much, even for Stroppie in her fragile state. It prompted her to open one rheumy eye and attempt a bubbly, sickly sort of snort which sent Punda bucking and kicking backwards around the enclosure. Stroppie's eyes opened a fraction wider in sheer disbelief. As they began to focus a little better, a look of outrage crept into them, for Punda, nothing daunted, was advancing for another onslaught on the horn. And so it went on – the more determined Stroppie was that she would not have her horn chewed the more determined Punda was that it would be chewed, and in that moment Stroppie realized that it was necessary for her to live – and she did.

The second rhino newcomer was newborn and therefore presented no great difficulty. He had been rescued in the nick of time, because he was actually being stalked by a rangy Tsavo lioness when his rescuers appeared. We never knew what had made him an orphan, for there was no sign of his mother anywhere, so we assumed he had simply been abandoned. His name was really supposed to be 'Pushmi', because that is exactly what he did most of the time, but Ali's mispronunciation turned him into 'Hoshim'.

Meanwhile, Eleanor was also having difficulties with rhinos, both tame and wild. Firstly, a wild rhino which had been fast asleep in the long reeds bordering the swamp, suddenly sprang to its feet, and with a furious snort, attacked Stub, tossing her clean over a bush. Meanwhile, the other orphans scattered – all, that is, except Lollipa, who was quite accustomed to consorting with rhinos and had not yet learnt to differentiate between tame and wild. The next moment she was hooked in the stomach by the wild rhino as it crashed past her into the bush. Pandemonium broke loose. Stub, thoroughly enraged, recovered her feet, and went baldheaded for the first thing she

Without further ado he grabbed her button-like horn between his teeth and began to tug *Photo: Gerard Vienne*

Hoshim clambers over Raru

happened to see which, unhappily, turned out to be poor Ali, catching him a glancing blow on the bottom just as, with agility born of long experience, he launched himself at a nearby branch. There he hung, swinging backwards and forwards whilst Stub cavorted around underneath him, tossing her head ominously up and down.

Fortunately, Ali's dignity suffered more than his bottom, but his trousers hung in tattered shreds as he swung. Loud yells brought Eleanor striding over, for she had been well in the lead at the moment the wild rhino popped up like a jack-in-the-box, and had not witnessed the subsequent sequence of events. Summing up the situation in a flash she hurled Stub aside with one swipe of her mighty trunk. Then, reaching up with her trunk, she tenderly began to investigate Ali, rumbling as though to reassure and console him until he plucked up the courage to let go his hold and drop to the ground. Ali was not, however, easily consoled, and when we arrived had a great deal to say about Stub and the hazards of herding creatures such as rhinos. Fortunately, Lollipa's wounds were not as serious as they might have been, and were nothing that attention and a little healing oil could not mend. As for Ali's, although there was not a mark to show for it all, and despite liberal applications of liniment and the acquisition of a new pair of trousers, his seemed to take forever to heal. For days afterwards he hobbled about, balefully bewailing his lot in life and treating us to lurid accounts of the wickedness of Stub, until David decided that something must be done about Stub. Ali, clicking his tongue and nodding his head vigorously, agreed wholeheartedly that something would.

This involved further adjustments to Eleanor's herd. Stub was sent first to the Nairobi National Park Orphanage, there to make the acquaintance of a captive male rhino. It was planned to introduce the pair of them into the newly created Elgon National Park on the Ugandan border which needed to be repopulated with rhinos.

Eventually, Stroppie and Punda were promoted to Eleanor's herd. Their introduction went far more peacefully than that of Stub. Stroppie had grown into a dreamy, placid, long-suffering sort of rhino, while Punda had eyes for Stroppie alone. Lollipa, the only other orphan which might have contested Punda's relationship with Stroppie, was far too stiff and sore to bother.

Stroppie and Punda, having grown up together, were inseparable. Evidently Punda looked upon himself as a rhino. His love for Stroppie was boundless, but it was a blindly possessive and jealous sort of love that was resentful of any other animal taking any notice of Stroppie. Punda seemed to think that Stroppie was his personal property. He subjected her to constant displays of his affection, which I am sure she could have done without, nipping her ears and tail, her heels and even her horn. After that initial battle of wills, and all

Stroppie and Punda were inseparable

resistance having been eroded over the years by Punda's incredible persistence, Stroppie had learnt that it was easier to submit. In time she became quite oblivious of his attentions, moving in a world of her own in which most things seemed to pass her by.

Punda resented Stroppie receiving attention from any other quarter

Quite understandably, Eleanor at first had serious reservations about Punda being in her herd; nor was she exactly elated at the prospect of yet another rhino, but eventually she accepted them with good grace. In time she even became quite fond of Stroppie, engaging her in gentle games, pinning down

In time Eleanor became quite fond of Stroppie . . . engaging her in gentle games

Stroppie's head with a huge foot in order to try and provoke some sort of reaction, sometimes even lying down, as Samson had done with Rufus, warding off Stroppie's onslaughts with a wave of her trunk. Thus, life reverted to its old routine and peace was restored among the orphans.

Raru

The rainy season in Tsavo is always a joyous time. It is heralded by a curious hush, as though every living thing waits in breathless anticipation: even the restless soughing of the wind in the trees and the singing of the birds is stilled. Great billowing clouds like huge wads of cotton-wool fill the sky, becoming darker by the moment until their inky bases trail in ragged wisps, and the sun beats down with heat-laden humidity, drawing beads of perspiration from every pore. It is then that the nyika seems almost to crackle with desiccated brittleness, and the Rangers say *Nchi na waka*, which means 'the land is on fire'. Eventually, amidst rolls of thunder that rend the heavens and blinding flashes of forked lightning, the great black clouds release their precious load. First in huge, slow drops that splash against the parched earth in puffs of powdery dust, then rapidly fuse together into a torrent that turns rivulets into streams, and streams into boiling, raging red torrents pouring down the many luggas to the main arterial rivers, and thence to the sea.

Almost overnight the nyika bursts forth with renewed life and is transformed into a lush wonderland, painted by a profusion of tiny wild flowers of every hue that nod and bob in the breeze and are constantly attended by scores of gaily coloured butterflies. Frogs awake from their long dry weather slumber and express their joy in an amphibian chorus that resounds through the night, the tinkling and bubbling and gurgling of the tiny tree frogs intermingling with the monotonous croak of the toads and the giant booming sound of the great bullfrogs. Fireflies dance like fairy torches, while overnight the rain-filled waterholes are hung with the frothy white glutinous baubles of breeding treefrogs, hitched to every available overhanging twig so that the infant tadpoles can plop safely into the water once they have hatched.

It is always like this in Tsavo. The rains bring renewed energy and life to the nyika. Millions of birds chorus a happy song, as they busily weave their intricate nests in trees that have burst with buds of green; all the shades of green in nature's palate: incredibly beautiful in lush and dramatic contrast with the autumnal tones of the dry season.

Every living thing was affected by this fever: even we were not immune – nor were our orphans. With the first drops of rain everyone went wild, doing all sorts of things that one would normally never do, like dancing in the rain, jumping into frothy, muddy pools, splashing and pelting one another with balls of sticky, red mud, or hauling out the old rubber dingy to race down the rapids of the swollen Voi river – always a favourite sport. Every year, the first drops of rain seemed to wash away our cares and problems, and we became irresponsible children all over again. Even David, who seemed more susceptible to 'wet weather fever' than all of us together, dreamed up all sorts of outrageous pranks that delighted the children and nearly drove me to distraction.

Our four-legged family behaved in the same way. Garden orphans raced around as though they had gone berserk; antelopes leapt into the air with bobbing white tails; guineafowl dashed about after one another; while Mr Honk, the peacock, screeched approval at the top of his raucous voice as he fanned out his magnificent tail for all to admire.

The one person who did not appreciate wet-weather fever, or feel much like partaking in the fun, was Ali, for when his large charges were infected by the fever, life for him was fraught with hazard. One could easily be run over by an elephant, for instance, tossed by a playful rhino, kicked by an ostrich or, worse still, horned by one of the buffaloes, not to mention what Punda might unexpectedly do. So, when the first drops of rain began to fall, Ali wisely stationed himself in the nearest and tallest tree until the excitement had subsided.

Eleanor, now a teenager, was beginning to feel the age-old tug of the wild. At this happy time, more than any other, she missed the companionship of other elephants, for during the wet season the wild elephants massed into large, sprawling concentrations to enjoy the festive period together. Whenever she happened to meet up with wild herds as she wandered in the nyika, Eleanor was clearly torn by divided loyalties. On the one hand she yearned for her own kind; on the other she was mindful of her responsibilities to the other members of her herd. She could not bring herself to take the final step and sever those human links forged by a shared life, love and friendship over many years. At the end of each day, when Ali's voice rang out to call her back, she was often reluctant to obey, pausing to swing round and gaze longingly from whence the sound of other elephants came, her trunk raised to savour that familiar scent, and her ears out to catch the fading voices of her own kind. Then, suddenly, almost purposefully, she would quicken her pace as though forcing herself to step back into the world of men. We sympathized deeply with Eleanor, for we understood her dilemma and knew the capacity of her large heart and the depth of the feeling that welled within it. She, more than any other elephant we had ever known, was an elephant of two worlds, faithful and loyal despite the strong dictates of instinct. And then she found Raru.

Eleanor was mindful of her responsibilities to other members of her herd

The instant Raru saw Eleanor he attached himself firmly to her side

We never knew how Raru came to be wandering alone in the wilderness, or what misfortune had made him an orphan. But there he was, lonely and forlorn – just as Eleanor had been fifteen years before. The instant he saw Eleanor he rushed up to her with a joyful trumpet, and attached himself firmly to her side. He was only about two years old.

Eleanor loved him immediately, taking him straight to her heart, rumbling her delight and caressing him with her trunk as lovingly and tenderly as any mother. And so, from that moment on, he belonged, following her everywhere as though he had been doing it all his life, even walking back with her straight into the stockades like a seasoned veteran, where he tucked in to the food provided for the night. Nor did the presence of Ali upset him unduly. His trust in Eleanor was complete: if she was all right, then so was he.

We had intended to name him 'Ndara', after the brooding massif overlooking the place where he had been found, but as had previously happened with 'Pushmi', Ali's mispronunciation soon converted it into 'Raru' – and Raru he has been ever since.

From the moment he arrived, it was obvious that in Raru we had an elephant that was quite different from most other

elephants. He was a born clown and show-off with a definite sense of mischief. Cocky, confident and bold, he would think up all sorts of antics in order to focus attention upon himself whenever anyone was watching. He would balance a bunch of grass on his head, or snatch Ali's battered old hat and place it on top of his own head, sit down on his haunches, roll his eyes round and wave his trunk in the air, or stand on his head and waggle his tail. If nobody else was present, Ali could always be counted on to provide an appreciative audience, for he always laughed as loudly as ten ordinary people, roaring with mirth every time Raru did anything that was even a little unusual.

Ali kept an eye on the orphans, usually from the vantage point of a tree

Now that she had Raru, Eleanor was much happier and more settled. She took him with her to meet the wild elephants, intermingling freely with them at waterholes whilst Ali kept a beady eye on the proceedings from a safe distance. Eleanor and Raru would wander away from the other members of the orphan herd to join the wild elephants, leaving Punda, Stroppie and the others with Ali.

Raru loved these excursions. Even when Eleanor sometimes hung back rather shyly, as she often did when confronted by a big bull, Raru was not in the least bit daunted. He took all sorts of liberties not usually allowed to young bulls, such as challenging an elephant three times his size to a sparring match, or jumping the hierarchy queue at watering places, or even forcing his way among strange cows in search of a suitable playmate. We were surprised that he did not take the opportunity of attaching himself to a more mature cow, but his devotion to Eleanor never wavered, and although he loved to play and fraternize with other elephants, and especially to show off, he always came back to Eleanor and was invariably in the column that filed back to the stockades each evening. Obviously, as far as he was concerned, Eleanor was now his new mother, and she fitted the part very adequately. Hence his loyalty lay with her.

It was not long before Raru found in Punda a kindred spirit as mischievous as he, and one on which he could work out his devilishness. This certainly relieved the pressure on Stroppie and diverted Punda for much of the time. Raru and Punda would spend hours on end tussling with one another, and in time a special relationship blossomed between them. The

Raru and Punda spent hours tussling with each other

Punda found Raru's tail irresistible

games they played were often pretty rough on poor Punda, who began to resemble a battle-scarred war horse, with tusk marks and ugly weals marring the beauty of his striped coat – though nothing seemed to deter him. With his usual incredible tenacity he persevered as though he had the hide of an elephant or a rhino, and seemed to enjoy every moment of it too.

Raru devised all sorts of ingenious tactics to corner Punda, for Punda was an elusive creature when it actually came to being caught. Reversing up to him, Raru would present his tail as an offering, and this to Punda was irresistible. Then, once Punda was hooked with Raru's tail clamped firmly in his teeth, Raru would slowly but surely steer him backwards towards a suitable bank or tree, kicking out first with one leg and then the other to keep Punda fully occupied until in position. Having successfully manoeuvred him against the bank or tree, Raru would spin round to pin him down and catch him with his trunk.

Tusks were another thing that Punda found irresistible. Feet had the same effect, for as far as he was concerned they were there only to be nipped. Often Punda would find himself caught with his neck wedged between Raru's tusks like a yoke, or with his head clamped in a vice-like grip beneath Raru's chin. The combinations were endless.

For Stroppie's sake we were very glad that Raru diverted Punda for so much of the time, and were surprised that the integration of little Hoshim and his baby buffalo friend into Eleanor's band went off so smoothly despite Punda's presence. Although Hoshim was a much more dynamic character than the dreamy Stroppie, Eleanor was by then more kindly disposed towards rhinos. As Hoshim – now a boisterous toddler of eight months – was beginning to disturb the tranquillity of the garden orphans' lives we decided to promote him at an earlier age than usual.

We chose an afternoon when Eleanor was bringing her herd back to the stockades, intercepting them on the way with Hoshim and his buffalo friend. Hoshim had been used to spending the mornings in the chicken run-cum-orchard at the back of the house, where he had been put to give the garden orphans some respite. There he enjoyed being tickled by the chickens, who clambered all over him as he slept to pick off ticks. Slowly, his eyes would close and his tail would rise as he was reduced to a state of blissful stupor, and the more the chickens jumped about on top of him and pecked at him the better he seemed to like it. So, when Eleanor marched up to him and began investigating him with her trunk, he collapsed like a pack of cards lifting up his leg to be tickled, while his little buffalo friend, who was not nearly so trusting, hovered about nervously on the periphery, prepared for instant flight. As for Hoshim, anything that tickled was divine, and Eleanor tickled far better than the chickens. He was not in the least bit awed either by her size or by the fact that he was surrounded by a curious throng of assorted spectators. He simply lay there with his eyes closed, exposing his tummy and sighing contentedly.

Eleanor obviously approved. She decided to have a game with him similar to the one she periodically played with Stroppie. Very gently she placed an enormous foot on top of Hoshim's head, exerting a little pressure which did not please him at all. His reaction was immediate, and made us all roar

Very gently Eleanor placed an enormous foot on Hoshim's head

with laughter: leaping to his feet with the whites of his eyes showing angrily, he spun round with a terrific snort, tossing his head up and down as he confronted her. Eleanor towered over him, her ears outspread like two gigantic sails. At first she was somewhat startled and then, hesitantly, out came her trunk again. Feeling the tickling starting once more, Hoshim rolled over on his back to make the most of it.

We were astonished at his command of the situation. Even Punda seemed slightly awed, for when he approached to make advances, little Hoshim rebuffed him so violently that he retreated and turned his attentions instead to Stroppie. As for Stroppie, she simply plodded on up the hill without even noticing that there had been any additions to the herd. Little Hoshim and his buffalo friend still had to have their bottle, or rather, gallon can, of milk twice a day, before they left the stockades in the morning and when they came back at night. I was usually there to perform this function, which fascinated Eleanor. Attracted by the sucking sounds, she always came over to supervise the proceedings, caressing both babies with her trunk as they fed, and rumbling softly to me and to them to express her affection.

Bukanezi

The price of ivory suddenly soared on the world's markets. So, after the quiet decade of the sixties, free from poaching, Tsavo's elephants once again found themselves under pressure as poachers moved into the Tsavo Park intent on making overnight fortunes from the sale of illegal ivory.

The Field Force had fallen below strength. David needed more men – those rugged men from Eleanor's homeland – so a recruiting expedition was hurriedly mounted. I remember the day the lorries returned with their fearsome looking passengers aboard, all chanting in unison.

On this particular day, the Big Orphans, including our handsome ostriches, now in full plumage, were going about their business close to the headquarters when the lorries swung into view. Immediately, the singing stopped: it was replaced by an ominous silence as every eye swivelled round and became firmly fixed on the unsuspecting ostriches. To these unsophisticated tribesmen an ostrich represents wealth: its feathers and skin, even pieces of its egg-shell are of value for personal decoration and tribal ritual. For them to have an ostrich almost within their grasp was like the fulfilment of a dream, for these birds were rare in their homelands.

The next moment, to our astonishment, warriors were leaping off the back of the moving vehicles and tearing after our ostriches, who wisely lost no time in getting away. Eleanor, sensing the general excitement, joined in the chase behind the new recruits, who by now were so bent on pursuing the ostriches that they were quite oblivious to all else. But not for long, for the scream of an angry elephant and the thundering of large feet suddenly brought them to their senses.

Chaos followed. Distraught warriors shrieked, some of them stumbling and falling to the ground and frantically scrambling to their feet again. All lost dignity and some their loincloths. We expected Eleanor, by now thoroughly excited, either to hurl a warrior into the air or trample one underfoot. It was fortunate for them that she was so obedient, responding to the shouts of both David and Ali by halting abruptly in her tracks.

The Sergeant of the Field Force who, to impress the new intake, had put on his best uniform, was enraged. Despite

Elephants were once again under pressure from poachers *Photo: Bill Woodley*

barking out orders in his most military voice, nobody had taken the slightest notice. He felt that he had lost face before the rest of the staff. Later, however, he obviously got his own back on the recruits, for within only three months it was difficult to believe that that row of smart, disciplined Rangers parading before David were these same unruly warriors – except, that is, when the ostriches appeared on the parade ground and strolled provocatively about. Then, as before, every pair of eyes moved with them until a ferocious glare from the Sergeant rectified this lapse.

With the Field Force reinforced, poachers were caught in droves. We were nevertheless saddened, for we knew that for every one that was captured there were scores of others to take his place. The exorbitant profits from the sale of a single tusk made the risk of a meagre prison sentence well worthwhile. I wondered if people would ever buy ivory again if they knew with what suffering it was bought. Among every bundle of arrows recovered from captured poachers, there were always the shafts of many that had been discharged, each representing a dying elephant somewhere in the nyika. So many poisoned arrows were recovered that they almost filled an entire store. The ivory that the Rangers picked up in the bush filled another, representing millions of pounds and hundreds of elephants. Suddenly the nyika seemed to be cloaked in deep sadness. We, who lived there and were a part of the nyika, could sense it, and with each elephant that died we suffered too, for we felt that we had failed.

Somewhere on the eastern boundary of the Park, a baby elephant stood in the blazing sun beside the inert body of his dead mother. Ever since the arrow had struck her, many weeks before, his life had dramatically changed. Until then it had been happy and carefree; days filled with new adventures as he roamed the nyika in the company of his family; fun with other baby elephants at mudwallows and waterholes; and always there was the protective umbrella of his mother's great body above so he never felt afraid. But his mother had been unwell now for many weeks. She had been growing weaker every day, and he could sense the mounting agitation of the herd as they hung back with her, trying to encourage her to keep up. Then, one day, she stopped and could go no more. Beneath the shade of a tree she waited resignedly for the end.

The herd had waited with her for as long as they could but, driven by thirst and hunger, had finally been forced to abandon her. Like Eleanor fifteen years before, the little elephant must have watched their departure with alarm and known for the first time the cold chill of real fear, for he had never been left behind by the herd before. When, finally, she crashed to the ground, kicking feebly until she died, terror must have gripped him.

For four days and nights he had stood pressed close to her lifeless form, days when hunger and a raging thirst nearly drove him mad and he repeatedly went down on his knees, scuffling about searching desperately for his mother's breasts, and nights when hyaenas encircled him, avidly watching every movement with glowing, hungry eyes. Had his rescuers not arrived when they did, a smaller corpse would undoubtedly have joined that of its mother.

This little elephant arrived at the headquarters in the evening. While he was being unloaded, Eleanor, who was already in her stockade, was beside herself with excitement, roaring her frustration in between loving rumbles, and stretching her trunk to its fullest extent through the bars of her stockade in an attempt to reach him as he was being hustled along. Then, we simply opened the door and pushed him inside before standing back to witness the boundless compassion of an elephant.

Gently, she drew him to her, caressing and talking to him softly, coaxing him beneath her tummy where she knew he would feel protected and safe. Her manner was more gentle and tender than she had ever adopted to Raru, as though sensing his distress and sorrow. Now that he was standing beside her, we realized how slim were his chances of survival. Barely three feet tall, he was only about a year old, and would be unlikely to be able to use his trunk properly, still being almost entirely dependent upon his mother's milk for nourishment. From past experience, we knew the inadequacy of any form of artificial substitute for elephant's milk, for baby elephants are extremely sensitive to diet. Certainly, this was the one animal that had consistently defeated us, despite having over the years tried every conceivable sort of concoction imaginable. Each time, our efforts had ended in failure and heartbreak. We therefore regarded a calf under three feet in height as beyond hope, and felt that the only chance for this little elephant was to try to persuade him to take green food earlier than usual, but in this we knew he would need some assistance.

Ali was therefore given instructions that the orphans be herded nearby for a few days until the calf was stronger, and that as much tasty fodder as possible be gathered and fed to it by hand in order to ensure that it had enough to eat.

Meanwhile, Raru's excitement had been just as intense, although nobody had paid him much attention. He had been trying to clamber up the bars of his pen like a circus elephant. Then he tried to squeeze underneath, his legs scrabbling for

An elephant-sized bottle of gruel *Photo: Mel Corfield*

purchase, but this, too, was a forlorn hope. In the end, having been defeated by both top and bottom, he decided to try the gap in the middle. The next moment, ear-piercing squeals made everyone jump. Much to our alarm, we realized that he had managed to get his head jammed in between the two rails. He was heaving and pulling and struggling so violently, in between distraught outbursts, that I was certain his neck would break at any moment. We heaved and shoved and pushed and pulled, as did Eleanor, but alas, to no avail. Just as we were beginning to feel utterly desperate, out it popped, quite unexpectedly, like a cork from a bottle, and everyone, Raru included, ended up in a tangled heap at the far end of the stockade.

Our relief was immense, although being in a heap with an elephant is not the best place to find oneself, especially if that elephant happens to have worked itself up into a state. Fortunately, nobody was damaged irretrievably, so all ended well.

Ali chose the name for our latest acquisition. 'It must be *Bukanezi*', he said, which meant 'the weak one' in his tribal dialect. So Bukanezi it was.

All went well with little Bukanezi for about two weeks. He was pampered and petted and cherished by Eleanor, and protected, too, from the persecution of Punda and the pranks of Raru. Attached firmly to Eleanor's side he was perfectly safe. She even allowed him to suckle her, although, of course, she had no milk, but it gave him comfort nevertheless. Ali picked copious quantities of green food which was stuffed into his mouth, and which he obligingly chewed and swallowed, so generally he seemed to be doing well, and we began to feel more optimistic about his prospects of living.

Then, one morning, Ali came hurrying to the house with the disturbing news that Bukanezi had collapsed and died half way up the hill behind the stockades. We rushed up the hill, and sure enough, there he was, prostrate on the ground, and indeed looking very dead, except that a slight heaving of his side betrayed the fact that life still flickered feebly within him.

My heart bled for Eleanor. She loved Bukanezi more than anything else on earth, as much as if he really were her very own, even to the exclusion of Raru and ourselves. Perhaps it was because he was so very small and dependent and needed such a lot of loving when he first came in that she favoured him so. The discharge from her temporal glands poured down her face in two dark lines as evidence of her distress.

Again and again she knelt down and tried to lift him to his feet, and we could see from the marks on the ground that she had been struggling with him for a long time. Raru hovered about too, though without being of much assistance, for all he did was to trumpet and charge Punda and Stroppie each time Bukanezi collapsed, as though they were to blame.

I was sent by David to mix a bottle of glucose and gruel – an elephant-sized bottle, holding a gallon – while the lorry was despatched to the town of Voi to try and find some special delicacies such as oranges and sweet-potato tops. Ali meanwhile went bustling off to call the Rangers to help get Bukanezi back on his feet.

When they arrived, Eleanor seemed to understand exactly what everyone was trying to do. She knelt down on one side of Bukanezi, supporting him with her trunk and with her tusks beneath his body. When David and the Rangers heaved from the other side she heaved too, so that by the time I arrived back with the bottle, he was up, being supported in the standing position by Eleanor. We opened his mouth and forced in the big rubber teat. Much to our surprise, he seized hold of it with his tongue and began to gulp down the contents of the bottle with great hungry sucks. Off I went for a second helping, and this too was demolished. He would have taken a third had David not felt it would do more harm than good.

Then, very gradually, we eased him back down the hill, the Rangers and David propping up one side, and Eleanor, with the strength of twenty men, coping quite easily on the other. By the time we reached the stockades, the lorry had arrived with the delicacies from town.

What happened next really did astound us, for normally whenever oranges were produced Eleanor forgot her good manners and was overcome by greed, shoving and pushing and grabbing as many as she could cram into her trunk. But on this occasion she hung back and calmly stood by while David sliced the juicy oranges and fed them one by one to Bukanezi. When Raru tried to force his way in and help himself, she sent him packing with a swift slap of her trunk. There could be no doubt that on this occasion Eleanor realized Bukanezi's greater need. We were impressed and touched by her restraint and intelligence. Down went all the oranges followed by the sweet-potato tops, with Bukanezi becoming visibly stronger every minute. After only another couple of hours, he was able to stand unsupported, though his legs trembled violently.

From that moment on he never looked back, making him the smallest elephant we had so far been able to raise. The bottle of gruel became a regular feature and he loved it. Nor were there

Eleanor repeatedly tried to lift Bukanezi to his feet

Eleanor seemed to understand exactly what everyone was trying to do *Photo: Jeannie MacKinnon*

Eleanor prevented him from collapsing *Photo: Jeannie MacKinnon*

any serious ill effects, which surprised us. As days passed into weeks, Bukanezi continued to thrive until he resembled a little round tub.

Eleanor spoiled him terribly, even more than before, because I think she never forgot just how close she had been to losing him. She fussed over him incessantly, and it was not long before he began to realize his privileged position within the herd and take advantage of it. Wandering nonchalantly over to Raru, he would instigate a duel by prodding Raru rudely in the side with his tooth-pick tusks. When Raru retaliated and things were no longer going in his favour, all Bukanezi did was yell. Eleanor would come running to his rescue, punishing poor Raru, quite unjustly we thought. Bukanezi was as demanding and precocious as any over-indulged human child, but in Eleanor's eyes he could do no wrong.

Every evening, loud trumpets and screams would tell me that it was time for his bottle and that he did not like to be kept waiting. As soon as I rounded the corner of the orchard and he spotted the bottle, he would career down the hill like a train, emitting triumphant trumpetings all the way. Faster and ever faster he came until, at the last moment, I had to leap aside to avoid being run over. Screams of frustration would follow as the bottle passed him by. This inevitably brought Eleanor running down the hill, which was even more daunting, especially when she was followed by all the other orphans as was often the case. Then I would have to sprint back up the hill with the bottle to the sound of pounding feet drawing closer and closer behind me. We all reached the stockades together with me puffing like a grampus and my heart pounding against my ribs from the excitement of the race.

Feeding Bukanezi did not become easier as time went by. Eventually I decided that if I was not to die prematurely of a heart attack I must shed my responsibilities and leave Ali to deal with the evening onslaught. In no time at all, however, he, too, was complaining. In the end we compromised by preparing the bottle well in advance and hiding it in the food store to be on hand the instant Bukanezi appeared. But Eleanor knew how to open the latch of the store, so we had to appeal to David, who devised a special Eleanor-proof catch – an elaborate combination of bolts that needed lifting and dropping as well as pulling, with a large round nut on the end which had to be unscrewed before the door could be opened. Only then was the bottle secure. Matters were further complicated by the bottles for little Hoshim and his buffalo stablemate, for the sight of them having a bottle drove Bukanezi into a frenzy. In order to

Punda joined in with enthusiasm

keep the peace Hoshim and the buffalo had to be fed surreptitiously in the seclusion of their stable. Even so, with that incredible elephant intuition, Bukanezi knew what was going on: the mere thought was sufficient to provoke an outburst.

Raru did not really like being overshadowed by Bukanezi. Even though Punda was now taking up more and more of his time he resented the intrusion and never missed an opportunity of having a dig at Bukanezi when Eleanor was not looking. Punda joined in the Bukanezi-baiting with enthusiasm, and this did not endear him to Eleanor, so it was rather a vicious circle, and Eleanor's normally peaceful herd became much more turbulent. But it was amusing to watch these two rascals,

Raru and Punda, awaiting their chance to get at Bukanezi, while at the same time keeping a wary eye on Eleanor.

We despaired of Punda. His love of rhinos was becoming an obsession that was beginning to extend beyond Stroppie and Hoshim to embrace all rhinos – wild as well as tame. At least once a week, grinning from ear to ear, Ali would relate with great glee how Punda had chased after a wild rhino, and had even seized one by the tail. We knew that this practice was courting disaster, and we wished that Punda could be a little more discerning in his choice of what to chase.

Below the headquarters was a special waterhole that normally filled after good rain. It nestled in a hollow with the distant hills mirrored in its glassy surface. It was neither too deep nor too shallow. The red mud there was perfect for plastering, and the bush that surrounded it offered a tempting variety of vegetation. A tall stately baobab, probably more than a century old, cast a cooling shade by the water's edge, and farther back, a leafy tamarind tree shed tart fruits for a tasty elephant dessert. In fact, everything an elephant loved most was to be found at the Red Waterhole. It was one of Eleanor's favourite places.

It was here that she had her first encounter with lions, although she must often before have seen them at night from the safety of her stockade, for they prowled around regularly, complaining with low grunts, and sometimes loud roars at their inability to get in. On this particular day these lions were busy feeding on an elephant calf they had succeeded in killing the previous night, and Eleanor detected their presence long before she actually saw them. Up went her trunk, and then without hesitating a moment more she charged forward through the bushes, screaming with fury and routing the lions which scattered in all directions. Angry growls of lion protest intermingled with sounds of enraged elephant were enough for even the boldest to make themselves as scarce as possible: the other orphans and Ali vanished. On this occasion it consequently took a long time to regroup, everyone having fled. In his hurry Ali lost his precious teapot. He bemoaned its loss so loud and long that I weakened and provided another.

Eleanor hated lions, as indeed do all elephants, for lions pose a very real threat to elephant calves. The Tsavo lions in particular were seasoned baby elephant killers, and sometimes of older elephants, too. Two mean old males that habitually lurked in the thickets of the lower reaches of the Voi river – fortunately far beyond Eleanor's beat – perfected the technique of bringing down quite large adolescent elephants, even some the size of Eleanor. They selected their victims with extreme cunning, picking on young bulls whose bonds were never as strong as those of the cows, and who would therefore be less likely to come to one another's aid when attacked. Whereas cow elephants will stick together and assist each other when things go wrong, young bulls are not nearly so noble or so brave, and at the first sign of serious trouble it is usually a matter of each for himself with little thought for the plight of another. Furthermore, those two cunning killers picked their moment well, for they were always most active at the end of the long dry season when the elephants were weakened by lack of food and therefore much easier to bring down.

We were incredulous that a lion could, or would, tackle a three-quarter grown elephant, but there was no doubt about it. Careful analysis of the spoor on the ground, coupled with an appraisal of the wounds inflicted on the victims pieced together the grizzly sequence of events and clarified the tactics employed by these two unusual lions that enabled them to bring down an animal four times their size and twenty times their weight. One merely hung on to the end of the victim's trunk like a bulldog, while the other attacked from below, sinking its fangs deep into the elephant's throat, so that death came from suffocation. Because they had evolved this technique, and found elephants easier to catch than other more fleet-footed species, they selected them in preference to other animals. An added advantage was, of course, that a single elephant would supply sufficient meat to feed them for several weeks. Lions are lazy creatures and do not like exerting themselves unless absolutely necessary.

These two lions took a particularly heavy toll during the great drought of the early seventies, when below average rains fell for six consecutive years, reducing the already arid nyika almost to a desert.

Droughts have always been and always will be a feature of areas of marginal rainfall such as Tsavo: they are something ordained by nature which lies beyond the control of man. They may even be necessary in arid lands to trim populations of large mammals such as elephants to the carrying capacity of the land and permit the large-scale regeneration of the nyika's scrubby bush. Nothing is ever static in nature; it swings back and forth like the pendulum of a giant clock that gauges time not in tens but in hundreds of years. There were years when Eleanor's favourite waterhole and many others like it failed to fill, and large stretches of the nyika remained brown and barren. In Tsavo water was life and everything hinged on the rain. If it came in gentle, healing showers, evenly spaced and spread over the entire Park, then all was well, for there would be food and water enough. If it came in violent drenching deluges and ran off the earth's hardened surface down the luggas, taking with it the topsoil, more harm than good often resulted. If it did not come at all or fell so intermittently that the first flush of green withered before the plants could cast their seeds, then the elephants, with their gigantic appetites and inefficient digestions were the first to suffer. But never had they been in such dire straits as during those six long, harsh years of the great drought. They died in their hundreds.

During this time of great suffering and sorrow, baby elephants were being orphaned nearly every day as mothers succumbed to the effects of malnutrition. From Eleanor's large heart flowed the boundless sympathy and comfort that an elephant can give in such generous measure, and which she lavished on one drought-stricken orphan after another whenever they were brought back by the Field Force.

Unhappily, by the time most of them arrived back they were usually already too far gone to be saved, and the shock of capture proved fatal, but we, and Eleanor, always did our best just the same, and were kept very busy. However, when despite our efforts one baby elephant after another collapsed and died, Eleanor sorrowed so sorely that we became worried

Eleanor leads the orphans through the nyika

about subjecting her to so much mental anguish, for she was showing signs of becoming somewhat neurotic, and even began to lose condition. Quite obviously, she came to associate lying down with death, and because of this did not like to see any of her charges lying down, even for a rest. Dashing over immediately, she would haul them back on to their feet to satisfy herself that they were not about to die as had so many others recently. Consequently, after a while, all her charges began to suffer from lack of sleep, going about in a daze like zombies. This was particularly true of poor little Hoshim, who being the youngest needed to sleep more than the others.

Occasional waifs and strays that Eleanor happened to come across whilst wandering in the nyika were also collected and brought back home, so that we were never sure exactly how many elephants to expect for the night. The stockades were often over-crowded. Sitting on the verandah, it became our practice to carry out a 'head count' as the orphans filed back at the end of the day. There were nearly always 'extras' that Eleanor had taken under her wing during the course of her wanderings in the Voi river valley. Most of those she found were young bulls who were not in a critical condition. After a brief sojourn and a good feed, they returned to the wild, to Ali's intense relief, for they often gave him a rough time. Each evening he appeared at our back door with a long face to have his scratches, scrapes, bruises and bumps doctored, all the while treating us to a harrowing account of how he had escaped death by a hair's breadth, having nearly been pounded to pulp by one of Eleanor's latest waifs.

Then, one day, a drought victim – a little female – was brought in which, in due course, recovered to become the fourth permanent elephant member of Eleanor's herd. We named her Sobo after the lone boulder of that name which marks a point on the Galana river where she had been found standing beside the body of her dead mother.

Sobo

Sometimes, when an elephant is under great stress, it will draw on reserves of water from its stomach, spraying the liquid over itself to cool its burning body. Inserting its trunk far back into its throat it sucks up this fluid as an extreme means of regulating body temperature, for elephants do not sweat through their skin as do most other creatures. Instead they rely on a large body area and the gentle fanning of their huge ears to keep them cool. When the sun is at its zenith they either retreat under the shade of trees or bathe in mudwallows and waterholes. If in desperate straits, however, they can call on nature's built-in safety device, drawing out quite large quantities of stomach water if need be.

Little Sobo, standing disconsolately on the treeless plain beside the body of her dead mother had to resort to this device. She sprayed herself repeatedly from mid-morning, when she was first found, until mid-afternoon when the lorry finally arrived with the Rangers to capture and save her.

On arrival back at headquarters, Eleanor took charge as usual. Bukanezi, who viewed every newcomer with suspicion and hostility and as a possible rival for Eleanor's affections, was beside himself with jealousy. Perhaps he sensed that Sobo was a very real threat, being more likely to live than any of her predecessors. After a few days, when Sobo's survival was a certainty, he became so obsessed with jealousy as to be thoroughly objectionable, behaving like a spoilt child.

Apart from Bukanezi's discomfiture, Sobo's arrival was opportune, for it not only diverted Eleanor's attention and kept her well occupied but restored her self-confidence, so that little Hoshim managed to get more rest and began to look more like his old self.

Sobo was a little older than Bukanezi, a forlorn and rather lonely little elephant, obviously harbouring deep scars from all the sorrow and anguish she had witnessed in her short life. Many of her friends and relatives must have perished in that devastating drought, and finally her mother. Small wonder, therefore, that she always seemed so silent and detached. Nevertheless, she fed well, stuffing lucerne into her mouth as though determined never to be hungry again. Before long her

Sobo was so weak that she had to be lifted to her feet *Photo: Jeannie MacKinnon*

Strong arms load Sobo into the lorry *Photo: Jeannie MacKinnon*

gaunt appearance softened into a healthier and more rounded shape, and her skin took on the supple texture of wellbeing.

Bukanezi was not really out in the cold for long, although he made enough fuss about it. He soon wormed his way back into favour, mainly by bullying poor Sobo until she drew away and he was able to monopolize Eleanor again. But there could also be no doubt that Eleanor still favoured him above all the others.

Nothing would ever change that. Despite this, he never forgave Sobo for having usurped his privileged place, albeit only briefly, and he continued to seize every opportunity to poke and prod and be as nasty as he possibly could, which made Sobo even more retiring. She would trail behind the others, occupied with her own silent thoughts.

We felt very sorry for Sobo. Raru was nice to her, but in a domineering bullish sort of way; in any event he found Punda much more fun. Eleanor was of course kind to her, too, but this only invited the wrath of Bukanezi. Neither did Sobo care much for the rest of the herd, particularly for that rascal Punda, whom time had not improved. She was therefore a loner, yet in spite of this determined never to be really alone again. On one occasion when she inadvertently became separated from the others and was lost for a night, she turned up back at the

That rascal Punda

stockades first thing in the morning and rushed joyfully in as soon as the gates were opened. Everyone was very glad to see her back – except, of course, Bukanezi, who simply sulked and made his feeling very plain.

And then one day not long after Sobo's arrival another orphan arrived unexpectedly. This time Bukanezi's nose really was put out of joint and the ordered routine of the orphans thoroughly disrupted. The new arrival was much more of a rival for Bukanezi, for it was much smaller and more dependent, a milk-only elephant – one of those delicate little creatures which were almost beyond hope and which I always dreaded having to cope with. It shuffled around as soon as it was lowered from the back of the lorry, ravenously hunting for a teat to which it could attach itself.

My heart sank. But the only alternative to attempting to save it was to destroy it – something that we could not bring ourselves to do.

This time, rather than cope on my own as before, I decided to enlist the help of Eleanor. Baby elephants need constant love and attention, and who better to provide this vital ingredient than Eleanor. I would provide the food and Eleanor would do the rest.

Predictably, this did not please Bukanezi at all, particularly as he had to be transferred into Sobo's stockade to make room for the tiny newcomer. Glaring moodily at what was going on inside Eleanor's stockade, he gave vent to a constant stream of frustrated bellows, to which no one paid the slightest attention, the new calf having eclipsed all else for the time being.

Eleanor was thrilled. With her great trunk she tenderly gathered the tiny, hairy, newborn baby to her, rumbling endearments and fondling and reassuring – to Bukanezi's mounting fury. Then, pushing it beneath her forelegs, she coaxed it to suckle. This was the moment for which I had been patiently waiting, armed with a bottle full of milk. Scrambling beneath Eleanor's forelegs from the opposite side, I substituted the bottle for Eleanor every time the calf tried to suckle.

Being actually underneath an elephant is a strange feeling. Perhaps it would have been an uncomfortable one had I not

Eleanor and her charges outside the stockade

known Eleanor as well as I did. To be scrabbling about among those gigantic legs, dwarfed by an enormous grey body towering above with the cold tip of a wet trunk periodically shaking all over you is an awesome and unusual experience. Beneath Eleanor's massive chin I could see her two gleaming tusks, each as thick and as long as my arm. I was very conscious of the latent power within that huge frame, and in her trunk, which could easily have crushed the life from my body. Yet, never for a moment did I doubt her, for I knew that I was perfectly safe. Deep rumbles from her throat which, being close to my ear, vibrated through my body, constantly reassured me.

I knew that Eleanor would understand and was confident that I could count on her cooperation. Ours was a deep and abiding mutual trust cemented by many shared experiences over a long period of time. Much more cause for misgiving came from the infant for whose benefit all this was being done. Instead of showing appreciation he meted out hefty shoves whilst we were both underneath Eleanor's stomach, which sent me hurtling backwards, still clutching the bottle, to land in disarray amongst trunks, legs, and the mound of vegetation placed in the stockade as food for the night. Then Eleanor would feel me tenderly all over and even Raru, peering interestedly through the crossbars separating the two stockades, would reach out his trunk to touch me, careful not to squeeze his head in between the rails, needless to say, but anxious to find out that I was still all right. Even that tiny elephant's strength was a salutary reminder to be sure to cultivate the friendship of an elephant – something that we were constantly drumming into Ali and his various assistants.

In this way all went smoothly for a while, or relatively so. During the day when the baby accompanied the other orphans, its bottle, the milk mixture, plus all the paraphernalia necessary to reheat it, had to go along as well.

I had decided on this occasion to feed the calf on demand as an elephant mother would, so Ali was given his instructions, and it was decided that he and I should take shifts. Every time the calf disappeared underneath Eleanor looking as though it was searching for milk, one of us had to be sure to be there to substitute the bottle. This was rather an exhausting performance and one that Ali did not exactly relish.

Most of the trouble stemmed from the other members of Eleanor's gang. From Punda, for instance, who sank his teeth into our bottoms whenever we bent down, and Bukanezi who, at the sight of the bottle, kept throwing monumental tantrums. He was especially aggravated by the sight of another animal feeding underneath Eleanor, which is where he would like to have been. Little Hoshim and his buffalo friend were also convinced that the bottle should be for them: simply knowing it was always there tantalized both them and Bukanezi beyond measure. Having to contend with all these difficulties certainly did not make calf feeding any easier. Consequently, by the end of the day, both Ali and I were worn out. Dust-covered, battered and weary, I would fall into David's strong comforting arms and need a lot of consoling as I poured out the problems of the day which compared to his were of course infinitesimal. However, out of bad came some good, for Bukanezi's frustrated bellows resounded through the nyika rather like the school bell. This enabled David to plot our progress throughout the course of the day, so that whenever he had a moment to spare, he would arrive unexpectedly to offer assistance and advice and see how we were getting on.

We named the little calf 'Gulliver', because although he was very young he was surprisingly large for his age. He shuffled about on feet and legs that lacked coordination and seemed far too big and long for his tiny body; he also had the hairy wizened look of a little old gnome.

Eleanor adored him, as did little Sobo who took upon herself the duties of a fully fledged nanny. We were happy that Sobo was able to fulfil this role because she obviously derived enormous pleasure from doing so. It also gave her something other than her own grief to think about and someone else to live for. At the same time, we realized, of course, that if we were to lose this baby which, because he was so very young, was a very real possibility, she would be hurt all over again. Never could there have been such a conscientious and dutiful little nanny, nor one that was so besotted with her charge. But she behaved

Gulliver

very discreetly, for although loving him as she did she must have longed to be close to Gulliver all the time, she was nevertheless careful to keep her distance until she was actually needed, accepting Eleanor's priority without question. Then, when Eleanor wanted a break from being a full-time mother, Sobo was there in an instant to take charge of Gulliver. She was always there when she was needed, never when she was not.

Eventually Eleanor began to rely more and more on Sobo, content to trust Gulliver to her while she went for a wallow or felt the need to investigate something unusual in the nyika. The rapport between these two females in the handling of a calf was very touching as well as being interesting to watch, giving us a much better appreciation than we had ever had before of the depth of those ties that bind individual elephants together. And the more we saw of them the more we realized that these gentle giants are not so very different from ourselves in their relationships with one another. Never was Gulliver left unattended even for a moment, nor could we ever really understand the means by which Eleanor and Sobo communicated with one another. It could only have been through telepathy, for each knew instinctively what the other wanted to do. Eleanor would simply move off and, as though by magic, Sobo would be there to step into her place; or if Sobo wanted to drink or wallow, Eleanor would be there at Gulliver's side. Although Bukanezi did not of course like having to play second fiddle, there was not much that he could do about it. We were glad to see Sobo settle down and become happier and much more contented, to feel useful and needed and even stand up for herself against precocious Bukanezi. One thing only clouded it all – the suspicion that little Gulliver was becoming weaker and the knowledge that he might not be with us much longer.

Eleanor got into the habit of leaving Gulliver with Sobo every evening when the orphans were returning to the stockades, for she had to hurry in order to catch up with Raru, having discovered that he had recently taken to stealing the food from her bin. Once on the homeward stretch, Raru would stride out purposefully ahead of all the others, gaining on them until he was well in the lead by the time he reached the stockades. Then he would cunningly sneak first into Eleanor's pen, and then into Sobo and Bukanezi's, greedily stuffing trunkfuls of the others' share down his throat, carefully keeping an eye on the corner of the orchard fence from whence the others would appear. As soon as they came in sight, he would slip quietly back into his own pen, presenting a picture of innocence so that when they arrived he was busily rummaging around in his own food bin as though no evil thought had ever entered his head. But Eleanor was not fooled for long. Probably she could smell that Raru had been at her share, and certainly she would have seen; hence the necessity to leave Gulliver behind with Sobo so that she could make haste up the hill after Raru, and woe betide him if she caught him in her bin.

Raru had devised some other pranks, two of which had to be discouraged: raiding the orchard was catalogued as a serious offence, while hurling unsuspecting tourists against the fence surrounding the stockades was another.

At first we were puzzled about the bulge in the wire just opposite Raru's particular stockade, but the mystery was solved by the occasional clue that tourist drivers let slip as they shepherded their flock around the stockades in the evenings to view the orphans.

'Watch out for the blighter in the middle,' we heard them say, so we felt it wise to ascertain the exact nature of this latest prank, although we knew Raru well enough to realize that no malice was intended. So we positioned ourselves strategically; then all we had to do was watch.

There he was, behaving in an exemplary manner as the tourists moved from one stockade to the other, 'oohing' and 'aahing' in disbelief, and gingerly reaching their hands through the bars to touch a live elephant. And then as they approached Raru's stockade, a calculating gleam crept into his eyes. Cunningly he waited until they were just opposite the bulge, and then with a quick push in the stomach with his trunk, they were sent hurtling backwards, caught totally unawares, to bounce back and forth in the bulge. The surge of excitement that invariably followed was obviously the part that Raru liked most, for understandably the victims were considerably shaken, and had to be helped off the ground, dusted down and consoled by many sympathizers who clustered about in a huddle casting dark accusing glances over their shoulders at Raru. This was of course exactly what he wanted, being always eager for the limelight.

We had to laugh, but at the same time we realized that this habit had to be discouraged, for we could not very well allow paying visitors to the Park to be bounced against fences, however entertaining it might be for elephants. So, out came the cattle-prod, powered by two torch cells, the instrument used to reprimand offenders guilty of any serious breach of discipline. It was much more effective than anything else, because the alien sensation of an electric shock, however slight,

Never was Gulliver left unattended even for a moment

was something they had never encountered before and did not much like. Its application was made more sinister by the ominous buzzing – rather like the sound of an angry insect – when the prongs were depressed.

When David meted out justice the next time he 'bounced' a tourist, Raru's bellows were quite out of proportion to the pain he could have felt: anyone would have thought he was being murdered. The cattle-prod was as notorious as any headmaster's cane, and equally as dreaded.

At this juncture a Japanese television crew arrived in Tsavo, and Eleanor's services were once again requested for the screen. This time it was for a Japanese children's television series featuring the travels of a curious creature that looked like a cross between a teddy bear and Mickey Mouse, but was really a little Japanese man inside a bright green, fluffy, puppet-like costume. We were told that the series was very popular in Japan as well as being educational, for this teddy bear-cum-Mickey Mouse travelled the globe visiting all kinds of interesting places where he had many exciting experiences. Now he had come to Africa to visit the Tsavo National Park, and meeting an elephant was of course essential.

David was rather dubious about the wisdom of this visit, not sure how Eleanor and the other animals would react when suddenly confronted by such an apparition. We felt fairly certain that they would not like it very much, and had visions of it and the crew being flattened in one fell swoop. To avoid the risk of people being squashed, we decided that a trial run was first necessary with the elephants in their stockades, and that the rhinos must definitely be barred from the act.

To everyone's astonishment, neither Eleanor nor indeed any of the elephants was fooled for a single moment. They knew instantly that this was no weird animal but just a man, and without a moment's hesitation promptly stretched out their trunks to accept the bananas and other treats it had brought along as a peace offering. Little Gulliver was enchanted. He, especially, was fascinated by the cuddly, furry feel of the fluffy apparition, snuggling up to it with trust, pulling at its ears and blowing salivary bubbles through his trunk.

Once again we were amazed at the intuition of elephants and their incredible intelligence. It would have been a very different story, I am sure, had the Japanese children's idol met an animal like a rhino, for instance. I doubt whether even the dreamy Stroppie would have failed to respond violently. But because an elephant is so special, those children thousands of miles away at the other end of the earth were treated to a particularly exciting episode in the travels of their imitation animal – the day it made friends with elephants in Tsavo.

The Poaching

Poaching continued to escalate, and the task of trying to combat it had now become a full time occupation, so that many other national park matters had to fall into abeyance, for there was a limit to what one man could do in the course of a day. I saw little of David at that time. He drove himself mercilessly, and seethed inwardly, realizing better than anyone the threat that poaching on this scale posed to the very survival of Tsavo. Although he tried repeatedly to impress on the authorities the gravity of the situation and the necessity for strong measures before poaching got out of hand, political considerations invariably intervened and he was not allowed his way.

Every day from dawn to dusk, Saturdays and Sundays included, it was the same. David would be airborne by first light at the controls of his supercub aeroplane, overseeing the campaign against the poachers, spotting fresh carcasses and occasionally poachers, and directing the Field Force Rangers from the air, giving them the back-up they so badly needed. At intervals he would land his plane on a sandbank in the river or on a road to refuel hurriedly from a drum dumped in the bush, so that within minutes he could be back in the air. Anyone who has tried flying a supercub aircraft eight hours a day, day after day, whirling round and round in tight circles only a few hundred feet above the ground, will realize just how taxing this can be. Often David would not return until after dark. He would then have to land with the aid of hurricane lamps to mark the runway, the sun having already sunk below the horizon; and when he arrived home he was utterly spent.

The plane was invaluable in the war against poachers. By flying low over them he was able to pin them down long enough for the Rangers, who were directed to the spot by radio, to catch up, and because of this close control the success rate was spectacular. There was no doubt about it – the Tsavo East Field Force was recognized as being the most effective anti-poaching squad in Africa. More and more poachers, nevertheless, poured across the Park boundaries. The fact that once a year the Government usually declared an amnesty to commemorate Independence Day, or some other such auspicious occasion, did not exactly help. The bulging jails would be opened up, and out would come all the convicted poachers, some of them only having served a fraction of their time. Back into the bush they went to seek another fortune, and we would be faced with the task of having to catch them all over again.

Carcasses festooned the arid nyika, some with their tusks still intact. Opportunists then began to join the general scramble for wealth, sneaking over the boundary in the hopes of being able to pick up ivory lying in the bush from elephants that had died but had not been found either by the poachers or the Rangers. Two such opportunists learnt a lesson they are likely never to forget, for Eleanor, with the help of Ali, was instrumental in their arrest, and in the process left them so terrified that it would be surprising if they ever set foot in the bush again.

One day, down near the Voi river, Eleanor suddenly halted in her tracks to listen, spreading out her ears, and testing the wind for scent. Ali stopped too, for he was astute enough to realize that here in the nyika Eleanor called the tune with senses finely attuned to her wild surroundings; and since the teapot episode his mind had been very much on lions.

Vultures clustering in the trees pointed to the fact that something must be dead, and then a faint chopping sound became apparent from beyond a bend in the riverbed. Eleanor was obviously slightly agitated, and imparted her premonition to the other orphans, who hung back nervously waiting for her to make a decision. For a few moments more she stood there, swaying uncertainly backwards and forwards, before making up her mind and going ahead, with Ali tugging at her tail in an attempt to dissuade her. In this he failed, so carefully keeping her between him and the chopping, which grew louder as they approached, he decided he had better go along as well.

Silently they rounded the bend in the riverbed. There before them, totally unaware of their presence, were two men in the process of hacking the tusks from the skull of an elephant that had recently died. At the same time they were warily keeping an eye on a group of wild elephants led by a seasoned old matriarch which was advancing cautiously downwind from the opposite direction.

Meanwhile, Eleanor with gigantic silent strides and followed by Ali closed the remaining gap without the poachers being aware of their presence. Not until Eleanor was almost within trunk range, and Ali in his most commanding voice had bellowed *Simama!* – which, in Swahili, means 'Stand!' – did they realize what was happening, and then they nearly leapt clean out of their skins with shock. Swinging around, they found themselves face to face not just with a man, which because of what they were up to would have been bad enough, but with an elephant with enormous outspread ears standing beside the man and staring at them.

Axes dropped from hands that suddenly went limp, knees began to buckle and knock together, and eyes rolled around in terror from faces that had taken on a lighter hue.

'Witchcraft!' one muttered feebly. Ali, who was not normally the bravest of people, was so encouraged by this reaction that he suddenly found he had all the courage in the world. With a menacing look on his face, he began dancing up and down in front of the two poachers, waggling his stick ferociously in front of their noses.

'Yes!' he roared, 'and you had better *simama* or else I'll instruct my friends here to deal with you!' and with an airy wave of his arm, he obviously included amongst 'his friends' not only Eleanor, but the wild elephants, which by now were beginning to look as though they might become involved, too.

Jabbering incoherently with terror, the two men pleaded on their knees for restraint, while Ali, by now thoroughly enjoying himself, strutted up and down pretending to ponder their plea, which terrified them all the more.

Meanwhile, Eleanor was carefully investigating the carcass, concentrating mainly on the tusks, by running her trunk up and down their gleaming surface. Then, with one foot on the skull, she gripped each tusk in turn firmly in her trunk, and heaved and heaved with all her might until with a sickening crunch they were torn from their sockets. Then, holding each one up in turn, she waved it around for a few moments before carrying it away and flinging it into the bush, almost as though she fully realized that herein lay the cause of the persecution of

Eleanor at the Red Waterhole *Photo: Jeannie MacKinnon*

Eleanor carefully investigated the carcass

her kind; for when suffering stalks the nyika, all who live there know and feel it too – even Eleanor, who was sheltered and shielded from the sorrows that beset her wild brethren. With elephant intuition, she nevertheless knew that all was not as it should be in the nyika.

This was the first elephant carcass she had ever seen, other than that of her mother, which she probably would no longer remember clearly. We wondered how it was she knew that tusks could be pulled out, and what she was thinking when she did it. How can we gauge the depth of an elephant's intelligence, and what goes on inside that massive head? To credit animals with human reactions and thoughts is considered 'anthropomorphic' and unscientific, but anyone who knows and understands animals appreciates that their thinking is not really so very alien to ours.

By now Ali had magnanimously decided to spare his captives' lives. After trussing them up with his battered belt, he began to march them back to the point where he had left the other orphans. The wretched poachers found themselves not only amongst elephants, which was frightening enough, but worse still amongst rhinos too, with a zebra, ostriches and buffalo thrown in for good measure. They were more than ever convinced that the forces of evil were afoot. Trembling violently, I doubt whether they would have been able to take a step had it not been for Stroppie's horn behind them, which kept brushing up against them, galvanizing them into action.

Approaching the staff houses, the procession accumulated a tail of interested spectators who tagged along behind, and in the process were treated to a much embellished version of the capture by Ali. As it happened, more than a tail was acquired at this point, for somehow Punda managed to get his hoof wedged in a discarded milk tin which had been thoughtlessly tossed into the bush and, what is more, so firmly wedged that no amount of bucking and kicking would dislodge it. Eventually, he gave up even trying, and rejoined the column still with the milk tin jammed on his hoof.

Ali, being very proud of his achievement, was determined to hand over his two captives to David personally. This led to a shouting match developing between him and the Field Force Sergeant, who felt that all prisoners should be his responsibility, with the result that he and some of his underlings also joined the procession in order to carry on the controversy. It was this hullabaloo that attracted our attention.

'Good Heavens!' exclaimed David, who happened to be home earlier than usual that day, 'this looks intriguing!' It certainly was. Eleanor was heading the column as usual, with Raru and Bukanezi hard on her heels. Next came Ali, one hand clamped firmly to the shirt-collar of each of his two prisoners; Stroppie behind, much to the prisoners' continuing discomfort; then Punda, clanking noisily up the hill, the can still on his hoof. The Field Force Sergeant and his supporters came next, arguing and guffawing loudly at Punda's expense; while sandwiched between them and the ragtag and bobtail who were there out of sheer curiosity, were the three ostriches, the buffalo, little Hoshim, Gulliver and Sobo as the rearguard.

It was not a simple matter to extricate Punda's hoof from the tin, for anything to do with Punda's nether regions was fraught with dangerous possibilities. But finally, after a struggle and a great deal of trouble, we managed to remove it. This was not achieved without cost, however, for at least four bruised and battered Rangers then had to be placated and patched, having somewhere along the line come into contact with Punda's flailing back legs, including the hoof that had been reinforced with the tin!

Eleanor's anti-poaching activities did not end with the capture of the two poachers. Sometime later she made history by unearthing a hidden cache of ivory that had been buried beneath a bush on the Voi river. At first, Ali was puzzled as to why this particular spot should hold such fascination for her, and was even a little impatient that she should be holding up everyone else for so long as she carefully investigated the ground about with her trunk. Then, down on her knees she went and began to dig into the ground with her tusks. Suddenly, to Ali's amazement and glee, out came one tusk, then another and another, until there were about eight in all, for which he was later able to claim the usual cash reward for 'found ivory'. Thereafter, whenever Eleanor felt she wanted to investigate any particular place Ali was all agog, hovering around hoping desperately that she was about to enrich him further by unearthing another cache; but, sadly for Ali, she never did.

Flying near the Galana river one evening on his way home from the usual gruelling hunting of poachers, David spotted yet another wounded elephant, this time a lone bull – so sick that he could hardly place one foot in front of the other. Obviously, he had been like this for some time, for he was desperately thin and emaciated, eyes sunken in their sockets and great bones protruding painfully from beneath loose skin that was hanging in dry, flaking folds. After each agonizing step, he stood for a long time, feeling his front leg carefully with his trunk. Even from the air David could see that the leg was swollen to three times its normal size where an arrow-head had obviously buried itself deep in the flesh.

Landing on a nearby sandbank, David groaned inwardly – another wretched victim to be despatched by a merciful bullet – it happened daily now, and thoroughly sickened him.

Taking his .416 rifle from the back of the plane, he began wearily to approach the wounded bull. When only a short distance from it he stopped in his tracks, for there was something painfully familiar about this particular elephant.

Appalled, he whispered to himself, 'It's Samson!'

David had a great affection for this elephant dating back to

A merciful bullet ended Samson's suffering *Photo: Jeannie MacKinnon*

the day twenty-eight years before when Samson had been brought in as an abandoned calf to become the very first orphan of Tsavo. With firmness tempered with kindness and understanding, David had soon won his trust and affection. Since then there had been a particular and very special rapport between them. Having saved him from certain death as a small, helpless calf, then nurtured and cared for him throughout his formative years, and finally persuaded him to sever his human ties and return to the nyika, we felt that we had succeeded. Samson had always loved and trusted mankind. Now, it fell to David to end his life. In that instant, more than at any other time before or since, David felt that he had failed.

Memories passed fleetingly through his mind as he raised his gun, and in the brief moment before the shot to end his suffering finally rang out, Samson looked up. Was there a flicker of recognition in those great sad eyes the moment before he died? David hoped not, for he felt ashamed to be a human being. Deep depression and a great sadness flooded over him. There was good reason why all four-footed creatures feared, detested and rejected man; but this one was different, and it was tragic that Samson who had been saved by the compassion of man, and who trusted man, should have lived only to become a victim of his cruelty and greed.

I was not told of Samson's tragic end until later, although I suspected that something out of the ordinary must have happened to depress David so. Probably he did not want to speak of it, being a man of few words who never betrayed emotion. Perhaps it was as well, for had we known at the time, the grief of the family would have been overwhelming. Even now, I am still sorely moved whenever I think about Samson, and I know that David must have felt his death more keenly than I, for he had had to shoot an old friend.

The Reunion

The great drought finally ended, but the jubilation that marked its termination was marred by the decline of poor little Gulliver. While the wild elephants rapidly recovered and roamed about the nyika revelling in the bountiful banquet of lush vegetation that spread before them, there was one little elephant who could not recover, no matter how much it rained in Tsavo. All my attempts to find a suitable diet on which to raise a newborn elephant had failed yet again, and eventually little Gulliver had joined the long line of casualties in the tiny graveyard that nestled peacefully beneath the dappled shade of the large Melia tree. David had planted this tree as a tiny seed thirty years before, and now it stood, majestic and beautiful, a living testimony to the passing of the years.

Tears of helplessness mingled with those of anguish when Gulliver died. Although I had many times tried adjusting his diet, desperately groping for something that would suit a baby elephant's delicate digestion, he had nevertheless become progressively weaker. Finally he had given up the struggle and had died by Eleanor's side in the dead of night.

Eleanor was distraught, as were we all, but it was Sobo who as I had feared felt the loss of Gulliver most keenly. We had each in our particular way developed a special affection for him. The sense of loss was thus more acute, for the involvement had been more intense.

Eleanor was perhaps able to accept the finality of death more easily than Sobo, for she had seen it many times during the past few years. She was in any event older, wiser and more philosophical about such matters. The morning when Gulliver failed to get up as usual she was nevertheless frantic, having spent a long time trying to coax some response from him before she was satisfied that there was no hope. Then, purposefully and deliberately, she had turned her back on sorrow and without even a backward glance had hurried away. She led her band of assorted followers where the sparkling rain-filled pools were full and rippling with life and the nyika bursting into the age-old miracle of its wet-season glory, more breathtaking, awe-inspiring and beautiful than before because of its longer than usual slumber. Already a profusion of pure white blossoms carpeted the parched earth like a huge, protective bandage as lacy foliage of dark, rich green interspersed with paper-thin blooms covered the countryside in wild profusion. It was almost as though, in celebrating the end of those long years of duress, nature was determined to decorate the nyika with an extravagance previously unmatched. Eleanor and her followers were soon romping and playing and rushing about trumpeting, celebrating with exuberant joy the advent of the festive season – all, that is, except Sobo.

Sobo had refused to follow the others. While we prepared Gulliver's grave, she stood beside his lifeless form, and when we laid him gently in the ground, she came up and touched him lovingly with her trunk, reaching down in a last farewell which added to the emotion that accompanied Gulliver's simple burial rites. Even when the grave had been filled with the damp red earth of Tsavo and decorated with some of those pure white flowers, Sobo stood beside it for a long while before slowly turning and walking away.

Each evening after that, when the orphans filed back up the hill, she left the others opposite the little graveyard and came to pay her respects at Gulliver's grave, silently remembering, silently thinking – who knows what. . . . As the days passed, we were further saddened to see her retreat into silence and solitude, again falling far behind the remainder of the column, preoccupied in her old detached way whilst the other orphans enjoyed themselves. In fact, so exuberant were they that Ali began to complain, for he evidently spent more time than usual clinging to the branches of trees in order to avoid being press-ganged into their riotous games. They were *mbaya* – bad – he said, all except Sobo. Only she was *mzuri* – good – because she was so quiet.

To cheer us up, David suggested that we should pay a visit to Eleanor's favourite waterhole and, like her, should walk through the nyika, savouring the heady scent of warm, wet earth and new green growth, soaking up the silence and the peace that was a component of the solitude of a wild place.

To walk in the wilderness with David was a magical adventure, for each step was filled with enchantment. As we went along, he would point out all sorts of fascinating things we would never have seen on our own, telling us about them so that we felt we were in wonderland.

We approached the waterhole cautiously just before noon, mindful of which way the wind would blow our scent, to ensure that our presence would not be detected by any animals in the vicinity which would be put to instant flight by a whiff of human scent. Then, selecting a small tree under which to sit, and one that looked unlikely to be required by any animal, we settled ourselves down to wait and watch and see what the day would bring.

Angela always mapped out her escape route well in advance, ending up a tree, and then secretly hoped it would have to be put to use, for she loved nothing better than shinning up trees, particularly as she was better at it than anyone else, except perhaps her father. Jill, being a good deal older, preferred not to climb, while my ineptitude in this respect was well known and resulted in my always being teased unmercifully. I would see Angela and her father conspiring together with knowing smiles and nods and surreptitious glances my way, deriving obvious amusement from the prospect of Mummy having to take to the trees! Then they would try to convince me that it was important to have a practice, because, they said, they wanted to be quite sure that I would be all right, but David's sly wink always gave the game away. My attempts at climbing trees were legendary, for inevitably I ended stuck half way up, unable either to go up or down – much to the delight of my family, who stood about underneath the tree offering ridiculous suggestions until I was frantic. Then, having enjoyed themselves, they would help me down saying: 'Never mind! With a rhino after you, you'd be up in a flash!' Fortunately this was never put to the test, so it remained open to question.

It was quiet and peaceful at the waterhole. Listening to the silence was almost hypnotic, for in it was the whispered music of another world – the voices of other tiny beings as they went about their busy lives, each making a minute but important contribution to nature's giant jigsaw. A column of harvester ants hurried past laden with the spoils of a foraging expedition.

A brilliant red dragonfly hovered like a helicopter over a tiny puddle, soon to be joined by a shimmering cloud of yellow and white butterflies which settled on the ground in the shape of a flower. A ferocious little ant-lion kicked up puffs of powdery dust as it repaired the treacherous walls of its pit, before settling down to wait patiently at the bottom until some hapless insect inadvertently slipped down the slope. Beside me a trapdoor spider suddenly opened the camouflaged lid to its cobweb-lined den, took a hurried look around, and shut it again. Every blade of grass housed some miniature form of life, and every square inch of ground was full of interest.

It intrigued me to contemplate their purpose; to ponder how tirelessly they toiled for the common goal and how selfless were the sacrifices they made in the process. What sort of cares, I wondered, beset this quaint community. Contemplating just a few possibilities, including that posed by the presence of the ant-lion, reduced our own problems into insignificance. When things go wrong, I always find it helps to reflect in this way; to project one's thoughts beyond one's own problems and think about those confronting the inhabitants of the nyika.

A louder than usual splash stirred us from our reverie and brought us back to the world of animals. A small group of elephants had arrived. With great slurping sounds they slaked their thirst, and then began to spray water over themselves in a shower of silver droplets that caught the colours of the rainbow as they fell. A lone bull then emerged from the bushes with a clump of uprooted grass wobbling about on top of his head. As he strode along, jauntily swinging his trunk from side to side, we knew from the swashbuckling way he walked that it could only be Raru. The orphans had arrived, and while Eleanor and the others were preoccupied sampling the green feast spread before them, Raru, being more sociable, strode up and, as usual, without any preliminaries, began to give the wild elephants a cocky greeting. For a moment or two they stared back at him in mild surprise before simply ignoring him. After a while, he thrust his way in amongst them to select from their midst a suitable sparring partner, which he then proceeded to bully and push around, bumping into other members of the herd exactly as though he was one of them. The next moment, the matriarch of the herd swung into action, and out shot Raru with a loud bellow, as though jet-propelled, having been prodded with her tusks.

Little Hoshim was the next to cause a disturbance. He wanted only to sleep, and as the shadiest and most popular resting place at the waterhole was beneath the baobab tree, he plodded single-mindedly towards it, regardless of the fact that it was already occupied by some wild elephants. Forcing himself in amongst the maze of legs, quite oblivious of the startled looks from faces peering between huge outspread ears, he threw himself down in their midst and promptly fell asleep. Wild elephants, which were not accustomed to such behaviour from small strangers, promptly erupted in all directions, trumpeting their alarm. It took some time for them to settle down again, this time under the shade of the tamarind tree, leaving the baobab to Hoshim.

All was peaceful for a time. We could see Eleanor and the others feeding quietly in the bush nearby. Raru had recovered himself and was engrossed in a game with Punda. Then an excited trumpet made us look up. Much to our surprise, who should be charging towards the wild herd as fast as her legs could carry her, emitting happy squawks as she ran, but Sobo. Without hesitation she joined the group at the waterhole. Rumbles and roars and other happy sounds of greeting filled the air as Sobo was welcomed back with unrestrained joy.

We knew immediately that this particular herd must be part of her past, and that she must be well known to them; that here at Eleanor's waterhole she had found the remnants of her lost herd – those fortunate few which had been able to withstand both the great drought and the poachers, and that in amongst them were close relatives. The reunion was certainly ecstatic. Sobo was ringed by a forest of questioning trunks that fondled and felt and smelt tenderly, as loving rumbles and roars resounded in the nyika and echoed against the hills mirrored in the waterhole. For us it was a particularly gratifying and heartwarming occasion, for in that instant we knew that the ghost of Gulliver had been laid. Amongst loved ones of long ago

Wild elephants along the Voi River

Sobo would find herself anew, and the comfort and love that would enfold her from this moment on would heal the hurt in her heart. All of a sudden she had become a very different elephant from the morose, sorrowing character of yesterday as she wallowed and splashed and romped with carefree abandon amongst others that she loved, as though she had not a care in the world. Eventually, when the wild herd began to move away, she remained among them, passing quietly and contentedly from our world to theirs, until the nyika swallowed her and, suddenly, she had gone.

Tears welled in my eyes and the two children were crying quietly too. But they were tears more of happiness than of sorrow, although, of course, we were sorry to see Sobo go, and would worry about her in the months and years ahead, particularly in view of what had happened to Samson. Nevertheless, whatever the future would unfold, the moment was a joyous one and I was happy that we had been there to witness it. Sobo had never truly settled with us. This was largely due to Bukanezi's precociousness, and of course, her elephant memory. Although we would probably never see Sobo again, or know what became of her, we were richer for having known her and would be thinking of her often in the years ahead. In a silent prayer of supplication, I prayed fervently for her safety and wellbeing.

'And what of Eleanor?' I suddenly wondered with a start. She was huge now, nearly twenty years old – virtually a full-grown elephant. Standing at the water's edge, she could have been any wild matriarch, but for the motley group that surrounded her. She was at an age when she too should really be living wild and thinking about having a family of her own. What deep joy such an event would bring her – and us. She mixed often with the wild elephants and sooner or later would probably be mated with a wild bull, so the possibility was very real. Would she, I wondered, then choose as Sobo had done, or were the ties that bound her to us too strong to permit such a painless escape. Secretly, deep within my heart, although it contravened that sacred vow, I hoped that they were, for I would not know how to say goodbye to Eleanor and could not envisage life without her. The prospect of losing her, of releasing her to the risk of death by poisoned arrow, filled me with such horror that I could not bear even to think about it. And yet at the same time I knew that I would have no influence in this choice, for it would be hers and hers alone.

On a sudden impulse, I decided that day to make a silent wish at the Red Waterhole. When the time came, one wish turned to many, for what better wishing-well than this, and what better moment too, when the crossroads of the highway of life loomed ahead, forking in many different directions. Our path would be dictated by events and bureaucracy; that of the orphans by instinct, inclination and providence. The prospect of one day having to say a final farewell to Eleanor filled me with such foreboding that I picked up a small, flat pebble and tossed it into the middle of the pool. As the ripples spread out in widening rings, I began to wish. . . .

My first wish concerned Eleanor; that for many years to come she would remain with us where I knew she would be safe, shielded from the nyika's natural upheavals and protected from the horror of poaching. I wished that she would have many children under our surveillance so that we would be able to share in her joy and watch a baby elephant of two worlds grow up as did their gentle mother.

The next wish was abstract; I wished that all people could have the opportunity to know an elephant as we had known and loved Eleanor. If this were possible, no one would ever kill an elephant again or condone such killing, nor buy the ivory of their tusks which was yielded by such suffering.

I thought then of Gulliver, and wished that I had been able to unravel the mystery of how to hand-rear a very tiny elephant calf. Then he and many others like him would not have to perish, but enjoy the same second chance afforded to many others.

Next, I wished for Tsavo. That this giant Park, which had weathered many storms with David at the helm, would continue to endure when he could no longer be there to guide its destiny. Unquestionably, stormier and more turbulent times still lay ahead. The logistics alone of efficiently running that giant Park posed almost insurmountable problems, quite apart from those posed by the quirks of nature and the pitfalls of misguided policies formulated by desk-bound bureaucrats – those armchair 'experts' to whom everything sounded so simple. I wished that wisdom might prevail amongst those in whose power it lay to influence the course of events, so that what David had achieved during these past thirty years would not relapse into irretrievable ruin. I also wished that those who followed in our wake would find it in their hearts to care for the Park with the same tireless dedication, so that Tsavo would remain a corner of Africa where wild animals would be left to live out their lives in peace and security.

Finally, I wished that no matter what happened, and even if we left Tsavo we could always return again, for people have to change from time to time, just as do the seasons, and I realized that we could not live in Tsavo forever. But I wished we could always return again to gaze on those broad, sweeping horizons tinged with mauve; feel again the thrill of a truly wild place; find peace in the twisted trunk of an old nyika tree and savour the fragrance of that terracotta earth. Also to hear again the magic cry of a vulturine guineafowl, the hum of a tsetse and the low, loving rumble of an elephant at peace.

I was concentrating so much on this last wish that I failed to notice David's quiet approach, and started when he spoke.

'A pebble for your thoughts!' He smiled, holding in the palm of his hand an almost identical pebble to mine.

I told him something of my wishes and ended by saying, 'I wish we could live here forever.'

For a long time David was silent, with the Red Waterhole mirrored in his eyes as he gazed into the distance.

'You can cut an old tree down,' he said quietly, 'but you can't easily transplant it.' And with that he skimmed his pebble across the waters of the Red Waterhole, where it bounced once or twice and then sank slowly to join mine in those murky depths.

Epilogue

Since those prophetic words, the path that was allotted us took several unexpected and tortuous twists that we could never have envisaged. Certainly it has not always been as smooth or easy as we would have wished; rather has it been more like the stony trails trodden by elephants that wander through the nyika, winding their way over hills and valleys, rarely straight or smooth. But then, like the elephant trails, one cannot always expect the path of life to be entirely without its twists and turns and stony patches – and we had led an enchanted life. Many were the times in the months that followed when I had to resort to my silent contemplation of the daunting difficulties that during the course of life beset ants and elephants alike, so that I could find the strength and courage to turn the dark corners.

But some of those waterhole wishes came true, or nearly so, and about these I will tell you so that this story of Eleanor might have the sort of ending I would like.

Today, Eleanor is still in Tsavo. She is still as gentle and dependable, and continues to pace those well-worn nyika trails with Bukanezi at her side, even though he is now almost as big as she. Each evening she returns to the stockades. She remains an elephant of two worlds. What is more, and because I can still visit her occasionally, I never did have to say goodbye. Each time I return, she greets me with a moving tenderness that tells me more than any spoken word, and I can gauge the depth of that feeling because I know her so well.

A herd of wild elephants moves across the nyika
Photo: Mel Corfield

Raru remains the rascal he has always been, though perhaps a greater rascal now. He comes and goes and, to the immense relief of those now in charge, goes for longer periods than he stays. But, when he does come, he still sneaks into the orchard in the dead of night to gobble up whatever he can find. Although perhaps I should not say this, I must confess I would rather Raru had what was left of my orchard than almost any one else.

And the others? Well, each took a different turn at those crossroads; some led to green pastures and, as far as I know, lived happily ever after. Some, like Punda, did not have the sort of ending we would have liked. But because, for the animals, we want a special ending we can say this:

All the orphans had a longer and happier life than they would otherwise have had, had they not been afforded that second chance. We controlled the destiny of two of them – Stroppie and little Hoshim, our two rhino orphans – for they carried on their noses a commodity that was now fetching five times the price of gold. Because the poaching of rhino was becoming so serious, we dared not leave them behind when we were posted from Tsavo, so we arranged a sheltered and comfortable existence for them in semi-captivity on a private ranch in the highlands. Every day I give thanks that we made this choice, for the black rhinoceros is a seriously endangered species and very few remain alive in Kenya today.

And what of us – the humans in this story? Well, the ending for us was not what we would have wished, for we suffered a devastating loss similar to that known by many elephant families during the great drought when, soon after we had left Tsavo, David died very suddenly and very unexpectedly. So, for us life has changed and can never be quite the same again, but like that of the elephants, it must still go on. Although we may not know it, no doubt it is unfolding as it should, guided by destiny and some unseen hand whose purpose is often not clear to us.

I look out on wild country still, and can see the blue horizon where it meets the sky, although I live now in a more suburban type of wilderness, different from the nyika of Tsavo, but beautiful and peaceful just the same. The birds still sing and the lions sometimes roar; the sun continues to rise and set in a blaze of glory, and giraffe and other gentle creatures wander past my home throughout the course of the day, so we feel we are amongst friends. Sometimes in the dead of night, when the darkness is deepest, a rhino comes to drink at a tiny waterhole below my garden – not a red waterhole like those of Tsavo, but the next best thing. There are old friends too amongst the buffalo of the Nairobi National Park where I now live, for the herd originated from orphans of Tsavo and elsewhere. When the herd is sprawled across the plain below my house, I still enjoy the feeling that part of Tsavo is with me.

One last Red-Waterhole wish was granted me before we had to bid the nyika farewell. Eventually I unravelled the mystery of how to feed a very tiny elephant, so that others like Gulliver have lived since. This was perhaps a fitting triumph on which to end my sojourn in Tsavo. One wish yet remains to come true: that of being able to share with Eleanor the joy of her baby – this time without the responsibility of having to raise it myself.

I often wonder about elephants, and marvel that they have all the better qualities of man and very few of the bad. Though great in stature, girth and strength, these characteristics are tempered with gentleness and restraint. Despite such size and strength, no other member of the animal kingdom is as noble and intelligent, as peace-loving, placid and tolerant, or is capable of the deep devotion and unswerving loyalty of an elephant, and none has its tender heart. All the noble qualities of these gentle giants have been engraved upon my heart forever, and for me are summed up in just one word – the name ELEANOR.